FOUNDATIONS³
**The Practical Resource for
Total Dust & Material Control**

FOUNDATIONS³

*The Practical Resource for
Total Dust & Material Control*

by

R. Todd Swinderman, P.E.
Larry J. Goldbeck
&
Andrew D. Marti

Martin Engineering
Neponset, Illinois
U.S.A.

Application of the principles described in this book should be carefully evaluated to determine their suitability for a specific project.

Information presented in this volume is subject to modification without notice.

For assistance in the application of the information presented here on specific conveyors, consult Martin Engineering.

One Martin Place
Neponset, Illinois 61345-9766 USA
Phone 800-544-2947 or 309-594-2384
Fax 309-594-2432
E-Mail: martinone@martin-eng.com
www.martin-eng.com

ISBN 0-9717121-0-7

Library of Congress Catalog Card Number: 2002090139

Copyright © Martin Engineering 2002

All rights reserved. This publication may not be reproduced in any form without permission from Martin Engineering, Neponset, Illinois, USA.

Printed in the United States of America.

contents

Preface ... iv
Dedication ... v
Foreword: Total Material Control ... 1

1. Belt Conveyors ... 8
2. Belting for Conveyors ... 16
3. Splicing the Belt ... 28
4. Tail Pulleys and Transition Areas ... 36
5. Loading Chutes ... 46
6. Belt Support ... 60
7. Skirtboard ... 84
8. Wear Liners ... 92
9. Edge Sealing Systems ... 98
10. Control of Air Movement ... 108
11. Dust Suppression ... 114
12. Dust Collection ... 124
13. Belt Cleaning ... 132
14. Tail Protection Plows ... 158
15. Keeping the Belt on Track ... 162
16. Considerations for Specific Industries ... 182
17. Safety ... 192
18. Access ... 196
19. Maintenance ... 200
20. The Human Factor ... 206

Acknowledgments ... 210
References ... 211
Index ... 214

preface

"Your input may be the key to the 'next level' of methods to assure cleaner, safer, and more productive bulk material handling."

The title *FOUNDATIONS³ The Practical Resource for Total Dust and Material Control* comes from our belief that in order to have clean, safe, and productive bulk material handling, plants need to assure total control of material on their belt conveyors.

This volume is the third in the series of Martin Engineering's FOUNDATIONS Books. It is a sequel to *FOUNDATIONS² The Pyramid Approach For Dust And Spillage Control From Belt Conveyors*, published in 1997, which in turn was the follow-up to *Foundations: Principles of Design and Construction of Belt Conveyor Transfer Points*, published in 1991.

Those texts were the descendants of Martin Engineering's 1984 publication *Conveyor Transfer Stations: Problems and Solutions*.

Each book represents the continuing growth and evolution of practices for the control of material movement and the improvement of belt conveyor operations.

We have attempted to construct this book in a non-commercial manner. The reader will note resemblance to various products of Martin Engineering. The simple explanation is that these products have been designed to reflect our philosophy and the lessons we have learned in "real world" applications on belt conveyors. Both this philosophy and these lessons are now represented in this text.

Your input may be the key to the "next level" of methods to assure cleaner, safer, and more productive bulk material handling. We would welcome your comments.

RTS, LJG, & ADM

February 2002
Martin Engineering
Neponset, Illinois USA
www.martin-eng.com

dedication

Richard P. (Dick) Stahura
Pioneer, Spokesman, Raconteur, Inventor, Entertainer, Educator.

"...he pioneered the engineered approach to solving the problems of dust and spillage around conveyors."

Dick Stahura has spent over 50 years working to improve bulk material handling. In that time–most of it with Martin Engineering–he pioneered the engineered approach to solving the problems of dust and spillage around conveyors.

Dick has carried his message around the world, to mines, power plants, and anywhere "DURT"–his unique term for the fugitive material released from belt conveyors–presented the opportunity to improve efficiency and productivity.

Like its predecessors, this book has grown out of Dick's vision that the clean, safe, and productive handling of bulk materials could be a direct result of properly cleaning and sealing belt conveyors.

Dick Stahura became so well identified with these concepts that for a number of years Martin Engineering has used his caricature as a trademark for its conveyor products. This character's name–"TC"–is short for the catchphrase Dick has long advocated as the first step in improving conveyor performance: "THINK CLEAN®!"

As the industries handling bulk materials feel greater needs to reduce dust and waste, to improve efficiency and productivity, to assure total material control, we must all remember, as Dick Stahura taught us, to THINK CLEAN®!

Edwin H. Peterson
Chairman
Martin Engineering

Stahura Self-portrait

Martin "THINK CLEAN®" Character

foreword

Total Material Control

Figure 0.1

Accumulations of fugitive material are a common sight underneath belt conveyors.

"Fugitive material has been around plants since the conveyors were first turned on..."

Material escaping from conveyors is an everyday occurrence in many plants. **(Figure 0.1)** It arises as spillage and leakage from transfer points, carryback that has adhered to the belt past the discharge point and is then dropped off along the conveyor return, or as airborne dust that has been carried off the cargo by air currents and the forces of loading. Sometimes the nature of the problem of a given conveyor is discernible from the configuration of the pile of lost material. Carryback falls under the conveyor, spillage falls to the sides, dust falls all over everything, including components above the conveyor. However, many conveyors show all these symptoms, so it is more difficult to place the blame on one type of problem. **(Figure 0.2)**

Fugitive material has been around plants since the conveyors were first turned on, so its presence is accepted as a part of doing business. In fact, the maintenance or production employees who are regularly assigned may see cleaning duties as a form of "job security."

Other employees may regard the problem of material that has escaped from material handling systems with resignation. They recognize it as a mess and a hazard, but they have found no effective, practical, real-life systems to control it. So they have come to accept spillage and dust from leaky transfer points and other sources within the plant as part of the routine and unalterable course of events. The fugitive material becomes a sign that the plant is operating–"we're making money, so there's fugitive material."

At one time, pollution–whether from smokestacks or from conveyor transfer points–was seen as a sign of industrial might. But now these problems are recognized as an indication of mismanagement and waste. At the same time, this pollution offers an opportunity for improvements in efficiency and bottom line results.

Left unchecked, fugitive material represents an ever-increasing drain on a conveyor's (and hence a plant's) efficiency, productivity, and profitability. Material lost from the conveyor system costs the plant in a number of ways; the following are just a few.

Reduced Operating Efficiency

It has been said the most expensive material in any operation is the material spilled from the belt. At a clean plant, material is loaded onto a conveyor and then unloaded at the other end. The material is handled only once–when it is placed on the belt. This means high efficiency, because the plant has handled the material as few times as possible.

When material becomes fugitive, it has to be gathered up and re-introduced into the system. This adds additional handling; the redundancy increases the overhead and, therefore, the cost of production. Once it has been released, the material is more difficult (and more expensive) to recover than to place on the belt the first time.

If the fugitive material cannot be reclaimed, the inefficiency is increased more dramatically. In many places, basic materials such as limestone or sand that falls from the belt is classified as hazardous waste and must be disposed of at significant cost.

Increased Conveyor Maintenance Costs

The escape of material from a conveyor leads to any number of problems on the conveyor system itself. These problems increase maintenance expenses for service and the replacement of components.

The first and most visible added expense is the cost of cleanup. This includes the cost for personnel shoveling or vacuuming up material and returning it to the belt (**Figure 0.3**). In some plants this cleanup cost will include the equipment hours on wheeled loaders, "sucker" trucks, or other heavy equipment used to move large material piles. A factor that is harder to track, but that should be included, is the cost of other work not performed because personnel have their attention diverted to cleanup. Other maintenance activities may be delayed due to the need to clean up the belt line, with possibly catastrophic long-term results.

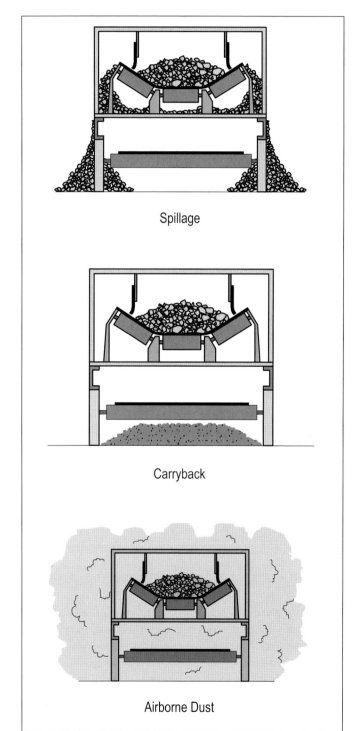

Figure 0.2

Three principal origins of fugitive material.

Figure 0.3

Cleaning up fugitive material increases labor costs.

Figure 0.4

Fugitive material can bury the load zone, resulting in idler failure.

As material escapes, it accumulates on the various conveyor components and other nearby equipment. Idlers can fail when clogged or buried under material. **(Figure 0.4)** No matter how good the seal on the bearing, fines will eventually migrate through the seal to the idler bearing. Once the bearings seize, the constant wear of the belt will cut through the idler shell with surprising rapidity. This leaves a razor-sharp edge on the seized roller that can badly damage the soft underside of the belt. **(Figure 0.5)** These "frozen" idlers and pulleys also increase the friction against the belt, dramatically increasing the power consumption of the conveyor.

Idlers facing fugitive material are sometimes replaced prematurely in advance of their expected wear life. This brings added costs in the form of equipment expenses, the labor costs for installation, and the cost of missed production and downtime required to perform the installations.

Another risk is that material buildup on the face of pulleys and idlers can cause the belt to run off center. **(Figure 0.6)** The belt can rub on the structure or even roll over on itself. If this condition is not noticed right away, the damage can destroy great lengths of valuable belting and even damage the structure. The accumulation of materials on rolling components can lead to significant belt training problems, which leads to damage to the belt and other equipment (as well as the risk of injury to personnel). This belt wander also creates interruptions in production, as the belt must be stopped, repaired, and retrained prior to resuming operations.

Figure 0.5

The motion of the belt across "frozen" idlers will wear rollers to sharpened edges.

What is particularly troubling about these problems is they become self-perpetuating: spillage leads to buildups on idlers, which leads to belt wander, which leads, in turn, to more spillage. Fugitive material is truly a vicious circle.

A particularly ugly circumstance is that fugitive material can create a problem and then hide the evidence. For example, accumulations of damp material around conveyor structures can accelerate corrosion, while at the same time making it difficult for plant personnel to observe the problem. **(Figure 0.7)** In a worst case senerio, this can lead to catastrophic damage.

Reduced Plant Safety

Industrial accidents are costly, both in terms of the health of personnel and of the volume and efficiency of production. The U.S. Bureau of Mines estimated the average cost of a fatality in an underground coal mine at $1.02 million. This figure includes the costs of medical expenses, worker compensation, accident investigation, loss of family income, and lost production value. The cost for accidents of less severity was calculated at $237,000 for permanent disability accidents and $5,000 for lost time accidents in underground mines.

Statistics from the Mine Safety and Health Administration indicate that roughly one-half of the accidents that occur around belt conveyors in mines are attributable to cleanup and repairs required by spillage and buildup. If fugitive material could be eliminated, the frequency at which personnel are exposed to these hazards would be significantly reduced.

Lowered Employee Morale

While the specific details of an individual's job have much to do with the amount of gratification received at work, the physical environment is also a significant influence on a worker's feelings toward his or her workplace.

A clean plant makes it easier to brag about one's place of work. As a result, employees have better morale. Workers with higher morale are more likely to be at work on time and to perform better in their assignments. People tend to feel proud if their place of work is a showplace, and they will work to keep it that way.

It is hard to feel proud about working at a plant that is perceived as dirty and inefficient by neighbors, friends, or especially by oneself.

Diminished Quality

Fugitive material can contaminate the plant, the process, and the finished product. Material reclaimed from the floor may have suffered degradation or contamination. Material can be deposited on sensitive equipment to adversely influence sensor readings or corrupt tightly controlled formulas.

In addition, fugitive material provides a negative factor in plant quality. Fugitive material sets a bad example for overall employee efforts. The most universal and basic tenet of many of the corporate "Total Quality" or quality improvement programs popular in recent years is that each portion of every job must be performed to meet the quality standard. Each employee's effort contributes to and reflects the entire quality effort. If employees see that a portion of the operation, such as a belt conveyor, is operating inefficiently–making a mess, contaminating the remainder of the plant with fugitive material–they will get used to accepting less than perfect performance. They may adopt a negative attitude and become lax or sloppy in the performance of their responsibilities.

Fugitive material provides a very visible example of the sloppy practices that the corporate quality programs are working to eliminate.

Heightened Scrutiny from Outside Groups

Fugitive material is a lightning rod; it presents an easy target. A billowing cloud of dust is going to draw the eye and the attention of concerned outsiders, including regulatory agencies and community groups. Accumulations of material under conveyors or on nearby roads, buildings, and equipment will send a message to governmental agencies and insurance companies alike. The message is that this plant is slack in its operations and merits some additional inspections or attention.

Figure 0.6

Material accumulation and idler failures can lead to mistracking.

Figure 0.7

Accumulations of fugitive material on the plant floor can create and hide structural problems.

If a plant is cited as dirty or unsafe, some regulatory agencies can mandate the operation be shut down until a problem is solved. Community groups can generate unpleasant exposure in the media and create confrontations at various permit hearings and other public gatherings.

A clean operation receives less unwanted attention from OSHA, EPA, and other regulatory agencies, and is less of a target for environmental action groups. This can provide cost savings by reducing agency fines, lawyer fees, and the need for community relations programs.

The Added Problems of Airborne Dust

In addition to the problems listed above, there are serious concerns with the airborne dusts that can escape from conveyor systems. And a greater problem is that while spillage is contained on the grounds of a plant, airborne dust particles are easily carried off-premises.

In the *Clean Air Act*, the US EPA is authorized to reduce the level of ambient particulate. Most bulk handling facilities are required to maintain respirable dust levels in enclosed areas below two milligrams per cubic meter. Underground mining operations may soon be required to meet levels of 1.0 mg/m3. Failure to comply with current and future air quality standards can result in stiff penalties from federal, state, and local regulatory agencies.

In addition, OSHA in the US has determined that airborne dust in and around equipment can result in hazardous working conditions. If OSHA or MSHA inspectors receive a complaint or an air sample that shows a health violation, it may lead to litigation.

Respirable dust–particles smaller than 10 microns in diameter–is not filtered out by the natural defenses of the human respiratory system and so penetrate deeply into the lungs and lead to serious health problems. These health issues might be seen in the workforce, and could occur in neighborhood residents.

One final problem that can arise from airborne dust is the risk of dust explosions. Any dust can concentrate to explosive levels within a confined space. The cost of one incident of this nature is far greater than the most expensive dust control program.

Regulatory Limits, ISO 14000, and the Environment

While no US regulatory agency has established specific limits on the amount of fugitive material allowed–the height of a pile beside the conveyor or the amount of carryback under an idler, for examples–there have been limits specified for quantities of airborne dust. These OSHA limits, called Permissible Exposure Limits (PELs) and Threshold Limit Values (TLVs), have been determined for some 600 regulated substances. These regulations specify the amount of dust expressed in millions of particles per cubic foot of air (millions of particles per cubic meter). It is industry's responsibility to comply with these standards or face penalties such as regulatory citations, legal action, increased insurance rates, or even jail time.

OSHA procedures note that inspectors should be aware of accumulations of dust on ceiling, walls, floors, and surfaces. The presence of this material serves as an alarm to the inspectors of the possibility of elevated quantities of airborne dust, which should then be measured with an approved sampling device.

Regulatory limits are different and will continue to differ from country to country. However, it seems safe to say the environmental regulations, including dust control, will continue to grow more restrictive around the world.

The continuing globalization of commerce promises more unified standards. Just as ISO 9000 has become a worldwide standard for quality procedures, the development of ISO 14000 will set an international agenda for an operation's impact on the environment. ISO 140000 prescribes voluntary guidelines and specifications for environmental management. As part of the program, it requires:

- Identification of a company's activities that have a significant impact on the environment.
- Training of all personnel whose work may significantly impact the environment.
- The development of an audit system to ensure that the program is properly implemented and maintained.

These guidelines could certainly be extended to include the fugitive material released from conveyors.

How a Little Material Turns into Big Problems?

Fugitive material escaping from transfer points presents a serious threat to the financial well-being of an operation. The obvious question is, how can it cost so much? A transfer point only spills a small fraction of the material that moves through it. But in the case of a transfer point on a conveyor that runs continuously, a little bit of material can quickly add up to a sizable amount.

If a belt that runs 10 hours per day releases just one shovelful–roughly 20 pounds (9 kg)–of material per hour, this loss totals two tons of material per month. (That translates to approximately 150 grams, or 1/3 of a pound of material per minute, lost as spillage, carryback, or windblown dust along the conveyor's entire length.)

Let us assume a conveyor has a demanding production schedule of 20 hours a day, 6 days a week, 50 weeks a year. If a transfer point on a conveyor with this operating schedule loses only 0.035 ounce (1 gram) of material–the contents of one of the packets of sweetener commonly found in restaurants–per minute, the total is 793 pounds (360 kg) per year.

In a more common circumstance, if this conveyor loses spillage of 0.035 ounces (one gram) for every ten feet (3 meters) of the conveyor skirtboard, and the transfer point has skirtboard of a length typical for this conveyor's operating speed, a speed of 1000 feet per minute (5 meters/second) results in an annual material loss of 6344 pounds (2880 kg) or more than 3 tons. **(Figure 0.8)**

This total is roughly 3 tons (2.8 metric tons) of material lost to the process; material that must be swept up from the plant floor and placed back onto the conveyor or disposed of in some appropriate manner.

In real life, fugitive material escapes from transfer points in quantities much greater than one gram per minute. Studies performed in Sweden and the United Kingdom have examined the real costs of fugitive material.

Research on the Cost of Fugitive Material

A 1989 report titled *The Cost to UK Industry of Dust, Mess and Spillage in Bulk Materials Handling Plants* examined eight plants in the United Kingdom that handled materials such as alumina, coke, limestone, cement, and china clay. This study, compiled for the Institution of Mechanical Engineers, established that industrial fugitive materials add costs amounting to a 1 percent loss of materials and 22 pence ($ 0.38) per ton of throughput. In short, for every ton of material carried on the conveyor, there is a 20-pound (10 kgs per metric ton) loss in material, as well as substantial additional overhead costs.

This overall cost was determined by adding four components together. Those components included:
1. The value of lost material (calculated at one percent of material).
2. The cost of labor devoted to cleaning up spillage, which averaged 6.9 pence ($0.12) per ton of throughput.
3. The cost of parts and labor for additional maintenance arising from spillage, which averaged 4.7 pence ($0.08) per ton of throughput.
4. Special costs peculiar to particular industries, such as the costs of reprocessing spillage, the cost of required medical checkups for personnel due to dusty environments, etc., representing 10.7 pence ($0.18) per ton of throughput.

Note: This loss includes fugitive material arising from problems, including spillage, conveyor belt carryback, and material windblown from stockpiles.

A similar study of 40 plants, performed for the Royal Institute of Technology in Sweden, estimated that material losses would represent two-tenths of 1 percent of the material handled and the overall added costs would reach nearly 7 Swedish Krona ($1.10) per ton.

It is interesting that in both these surveys it was actual material loss, not the parts and labor for cleanup and maintenance, which added the largest cost per transported ton. However, the indirect costs of using labor for time-consuming cleanup duties rather than for production are not included in the survey. Those figures would be difficult to calculate.

It is easier to calculate the actual costs for the disruption of a conveying system that, for example, lowers the amount of material processed in one day. If a belt runs 24 hours a day, each hour's production lost due to a belt outage can be calculated as the amount (and the market value) of material not delivered from the system's capacity. This affects the plant's revenues (and profits).

Many practical problems exist in accurately determining the amount of fugitive materials, their point of origin, and their economic impact.

The Economics of Material Control

The cost of systems to control fugitive material is usually considered three times during a conveyor's life. The first consideration is during system design; the second, at start-up; and the third, during ongoing operations.

During the design stage, the primary economic consideration is to keep the capital cost of the new conveyor system to a minimum. The detailed engineering of a transfer point is time-consuming and usually left until after the contract has been awarded. At that point, the system is designed by a junior engineer or draftsman to minimum established standards, unless the owner insists on particular specifications.

Annual Material Spillage		
Belt Speed FPM (m/s)	Skirtboard Length Per Side (Typical) Feet (Meters)	Material Loss lb (kg)
250 (1.3)	10 (3)	1586 (719)
500 (2.5)	20 (6)	3172 (1439)
750 (3.8)	30 (9)	4758 (2158)
1000 (5)	40 (12)	6344 (2880)

Based on 0.035 ounce (1 gram) of leakage per minute for every 10 feet (3 m) of skirtboard during annual operating schedule of 20-hours-per-day, 6-days-per-week, 50-weeks-per-year. Skirtboard length is typical for belts of specified speed.

Figure 0.8

Annual material spillage calculation, based on loss of 0.035 ounce (1 gram) per minute for every 10 feet (3 meters) of skirtboard.

It is often very difficult with new installations to predict the precise requirements for material control. In most cases only a guess can be made, based on experiences with similar materials on similar conveyors, indexed with "seat of the pants" engineering judgments. An axiom worth remembering is this: A decision that costs $1 to make at the planning stage typically costs $10 to change at the design stage or $100 to correct on the site. The lesson: it is better to plan for worst-case conditions than to try to shoehorn in additional equipment after the initial system has been found to be under designed.

The details of conveyor transfer points, such as the final design and placement of chute deflectors, are sometimes left to the start-up engineer. It may be advantageous to allow the suppliers of specialized systems to be responsible for the final (on-site) engineering, installation, and start up of their own equipment. This may add additional cost, but usually it is the most effective way to get correct installation and single-source responsibility for equipment performance.

The Importance of Record-Keeping

Much attention is paid to the engineering of key components of belt conveyors. But too often, the factors that will affect the reliability and efficiency of these expensive systems are ignored. The cost of fugitive material is one such factor.

Record-keeping on the subject of fugitive material is not part of the standard reporting done by operations or maintenance personnel. The amount of spillage, the frequency of occurrence, the maintenance materials consumed, and the labor costs are very rarely totaled to arrive at a true cost of fugitive material. Factors such as cleanup labor hours and frequency; the wear on conveyor skirting; the cost of idler replacement in price, labor, and downtime; and the extra energy consumed to overcome stubborn bearings as they start to seize due to accumulations of material should all be calculated to determine a true figure for the cost of fugitive material. Components whose service life may be degraded by fugitive material–such as idlers, pulley lagging, and the belt itself–should be examined so that their replacement and service cycles can be determined.

The measurement of fugitive material at transfer points is difficult. In an enclosed area, it is possible to use opacity measuring devices to judge the relative dust content of the air. For transfer points in the open, dust measurement is more challenging, although not impossible.

A basic technique is to clean a defined area and weigh or estimate the weight or volume of material cleaned and the time consumed in cleaning. Then, follow up with repeat cleanings after regular intervals of time. Whether this interval should be weekly, daily, or hourly will depend on plant conditions.

What will be more difficult to determine is the point of origin of the lost materials. Fugitive material can originate as conveyor carryback, belt wander, skirt seal leakage, loading surges, off-center loading, leakage through holes in chutework caused by corrosion or missing bolts, or even dust from floors above.

The individual making this study has to bear in mind the number of variables that may influence the results. This requires that the survey be conducted over a reasonable time frame and include most of the common operating conditions–such as environmental conditions, operating schedule, material moisture content, and other factors that create or complicate problems with fugitive material.

Record-keeping of amount of spillage and of the costs of labor, parts, and downtime associated with it should be a key part of the management information system for the operation of belt conveyors. Only when armed with such records covering a period of operation will an engineering study of fugitive material and recommendations for total material control seem reasonable.

For many conveyor systems, the costs associated with lost material will easily justify corrective measures. In most cases where adequate records have been kept, it has been shown that a very modest improvement in material control will rapidly repay its costs. The savings in labor expense alone will often offset the cost of the retrofit equipment installed in less than a year.

The Opportunity for Total Material Control

When the costs created by fugitive material are understood, it becomes obvious that the control of material at conveyors and transfer points can provide a major benefit for belt conveyors and to the operations that rely on these conveyors. But this control has proven difficult to achieve and more difficult to retain.

What is needed, then, is a planned and maintained approach to total material control. This is an opportunity to reduce costs and increase efficiency and profitability for many operations. This is the opportunity for total material control.

Total material control means that you keep material on the belt and within the system. You move material where you need it, in the condition you need it, at the flow rate you need it, without material loss or excess energy consumption, and without premature equipment failures or excessive maintenance costs. Total material control improves plant efficiency and reduces the cost of ownership.

The following is a program to achieve total material control for belt conveyors.

Chapter 1

Belt Conveyors

Belt conveyors have been used for decades to transport large quantities of materials over long distances. In comparison to haul trucks, wheelbarrows, or woven baskets perched on a human's head, belt conveyors have proven time and time again that they are a reliable and cost-effective method for material movement.

Belt conveyors can transport materials up steep inclines, around corners, over hills and valleys, across bodies of water, above ground, or below ground. Belt conveyors integrate well into other processes, such as crushing, screening, railcar loading and unloading, stock piling, and ship loading.

Belt conveyors have shown the ability to transport materials that vary from large, heavy, sharp-edged lumps to fine particles; from wet, sticky slurry to dry, dusty powders; from foundry sand to potato chips; and from tree-length logs to wood chips.

Of all the material handling systems, belt conveyors typically operate with the lowest transport cost per ton, the lowest maintenance cost per ton, the lowest power cost per ton, and the lowest labor cost per ton.

Note: this text discusses belt conveyors used to carry bulk solids–that is loose, unpackaged, more-or-less dry material. Belt conveyors that carry packages or units, while in some ways similar, are not included in this discussion.

Conveyors 101

In essence, a belt conveyor is a large rubber band stretched between two (or more) pulleys, traveling at a more-or-less consistent, more-or-less high rate of speed, carrying a quantity of materials. For many plants, the belt conveyor is the basic building block of bulk material handling. Complications arise as the line of travel becomes inclined or curved and when the conveyor must be fit into a sophisticated process or plant to meet material feed-rate requirements and other operational constraints.

A belt conveyor is a relatively simple piece of equipment. Its basic design is such that it will convey material under the most adverse conditions–overloaded, flooded with water, buried in fugitive material, or abused in any number of other ways. The difference, however, between a correctly engineered and maintained belt conveyor system and a dysfunctional system usually becomes apparent in the system's operating and maintenance costs.

Every solids-handling belt conveyor is composed of six major elements:
1. The belt, which forms the moving surface upon which material rides.
2. The belt support systems, which support the carrying and return strands.

"For many plants, the belt conveyor is the basic building block of bulk material handling."

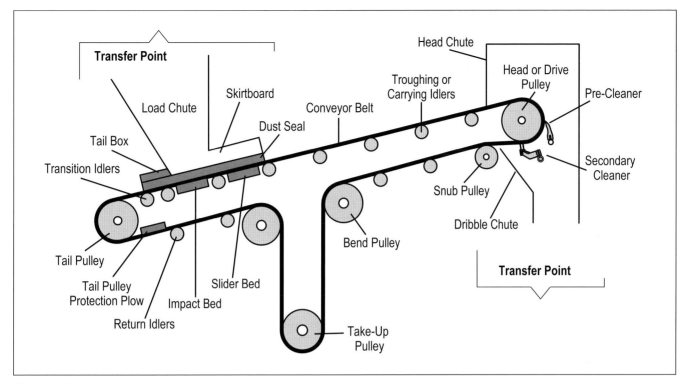

Figure 1.1

Conveyor component nomenclature and typical location.

3. The pulleys, which support and move the belt and control its tension.
4. The drive, which imparts power to one or more pulleys.
5. The structure, which supports and aligns the rolling components.
6. The enclosures, through which the belt's cargo is deposited or discharged.

Another part of every conveyor is the ancillary equipment installed to improve the belt's operation. This would include such components as take-ups, belt cleaners, tramp iron detectors, skirtboards and seals, plows, safety switches, dust suppression and collection systems, and weather protection systems.

Belt conveyors range in size from 12 inches (300 mm) to 120 inches (3000 mm) in belt width, and may be up to several miles long. Cargo capacity is only limited by the width and speed of the conveyor belt, with bulk conveyors often moving several thousand tons of material per hour, day in and day out. The amount of load can be varied with variable rate feeders or by controlling belt speed.

Components of a Belt Conveyor

Although each belt conveyor is somewhat different, they share many common components. **(Figure 1.1)**

A conveyor consists of a continuous rubber belt stretched between two pulleys. One end is called the tail end, and this is usually where the loading takes place. The other end is termed the head end. The conveyor usually discharges at the head end, but with the use of plows or trippers, may discharge anywhere along its length. Loading may also take place anywhere along the length of the conveyor belt, and multiple load zone conveyors are relatively common.

The belt is supported along the top (or carrying) side with flat or troughing rolls or idlers. Troughing rolls are used to increase the capacity of the conveyor belt. The belt is supported on the bottom (or return side) with flat or V-shaped return idlers.

The conveyor's drive motor is most often located at the head pulley, but may be located anywhere along the return side of the conveyor or even at the tail pulley.

The rolling components are mounted on and supported by the conveyor stringers.

The system incorporates mechanical or automatic tensioning devices or take-ups to make sure that the belt remains tight against the drive pulleys to reduce slippage and improve energy efficiency. The automatic tensioning device is often referred to as a gravity take-up or a counter weight.

Snub pulleys, used to increase the wrap of drive pulleys, and bend pulleys, used to turn the belt into a take-up pulley, are installed in the conveyor return side to ensure suitable tension of the belt against the drive pulleys and the automatic take-up.

The conveyor's loading end is called a load zone or a transfer point. The head end will probably consist of the

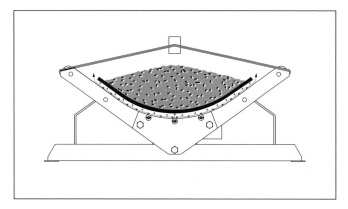

Figure 1.2
Air-supported conveyors replace carrying idlers with a film of air.

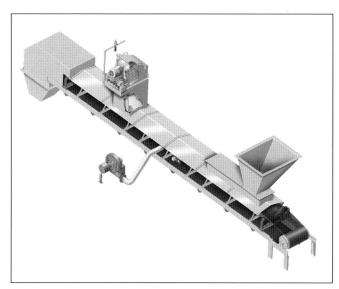

Figure 1.3
As they are totally enclosed, air-supported conveyors offer effective dust control.

Figure 1.4
Air-supported conveyors can be retrofit onto existing troughed-idler conveyors.

head pulley, belt cleaners, head chute enclosure, dribble chute, and discharge chute. The tail end consists of the load chute, tail pulley, belt support systems, wear liners, dust seals, entry seals, and exit seals. Depending on the conveyed material, dust collection or suppression systems may also be used in the transfer points at either end or along the run of the conveyor.

The discharge end should also be considered a transfer point, as in most cases one conveyor is used to feed a second conveyor, or another transportation device, storage system, or piece of process equipment.

Alternative Conveying Systems

While for most applications the conventional troughed idler conveyor dominates the market, there is a growing appreciation of alternative belt conveyors for special applications. These systems still use a belt to carry the load; but to accomplish a number of other goals, other components have been changed. This allows these conveyors to provide different capabilities or serve different applications.

These systems offer effective alternatives to conventional conveying when there are special requirements or space limitations. While each is unique, they still share many of the common benefits and problems associated with conventional conveyors. Without reference to trade names or proprietary information, the following discusses a number of these alternative conveying technologies.

Air-Supported Conveyors

Air-supported conveyors use conventional belts, pulleys, and drives; but in place of the carrying idlers, use a film of air to support the belt and cargo.
(**Figure 1.2**) A trough-shaped covered plenum that is fed with low-pressure air from a blower replaces the idlers. A series of spaced holes in the plenum allow the air to escape under the belt lifting, it off the plenum for the effect of a low-friction air bearing. A single, relatively small horsepower blower can support up to 600 feet (180 m) of conveyor.

One advantage of air-supported conveyors is that the system lends itself to effective dust control, as the belt is totally contained between the plenum and cover.
(**Figure 1.3**) The use of the air film to support the belt eliminates the idlers and skirt seals, thus eliminating maintenance on these items. Air-supported conveyors eliminate the undulating "roller coaster ride" provided by idlers, so they do not agitate or segregate the load. They are quiet in operation, often reducing noise levels by 15 dBa. They are also energy-efficient, using from 5 to 30 percent less total energy than a comparable conventional idler conveyor.

New designs for air-supported conveyor plenums allow them to be retrofit onto existing conveyor

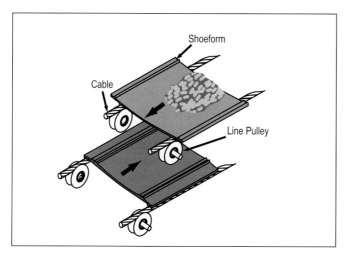

Figure 1.5
Cable belt conveyors use a wire rope to support and move the belt.

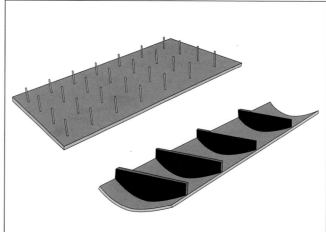

Figure 1.6
Belt cleats are available in a variety of configurations.

structures. (**Figure 1.4**) Some systems are modular, allowing them to be installed in sections rather than requiring a total conveyor replacement. Because of the energy savings and the elimination of cargo turbulence, the capacity of air-supported belt conveyors can often be upgraded from the conventional conveyor they replace by increasing the drive speed.

The primary disadvantages of air-supported conveyors are that they must be center-loaded to stay on track and they are limited to light-impact applications. Enclosing the return run is often not cost justified unless there is a special need. Air-supported conveyors must have well-engineered belt cleaning systems to keep the belt tracking properly on its return run and prevent the buildup of material on the bottom side of the belt, which could choke the air supply holes below the belt.

Typical applications for air-supported conveyors are where dust or contamination is an important consideration, such as in handling of grain or crushed coal. Many other materials have been successfully conveyed with excellent results on belts up to 72 inches (1800 mm) wide at speeds up to 1000 fpm (5 m/sec).

Cable Belt Conveyors

Cable belt conveyors utilize a special belt with high cross stiffness and tracks molded into the bottom of the belt's outer edges. (**Figure 1.5**) A pair of wire-rope cables fit into the tracks to support and guide the belt. The cables are strung on a series of special wire rope pulleys. The return run is carried on conventional return rollers. The loading and discharge resembles a conventional conveyor with free rotating pulleys that accommodate the tracks in the belt. The cables provide the tension and driving force for the belt.

The advantage of the cable belt is in the concept of separating the cable as the support and driver from the belt as the load-carrier. This allows for efficient energy use and long distance conveyors. Because it is the wire rope that is guided, the conveyor can make horizontal and vertical curves without problems. The load is not riding over idlers, so it avoids segregation and condensation during its journey.

The disadvantage of the cable belt conveyor comes in the design trade-off of how strongly the cable is gripped in the track and how much friction is provided to the belt at the loading and discharge points. Since the pulleys "free wheel," it is sometimes difficult to clean or seal the belt because the hardware needed to accomplish this can cause the belt to bunch up or stretch. In addition, weather can sometimes cause a loss of friction between the cable and the belt.

Common applications where cable belts can be very economically installed and operated are very long overland ore-handling conveyors.

Cleated Belts

Cleated belts are belts that have had large ribs, fins or chevrons attached to the belt surface. (**Figure 1.6**) These cleats can be vulcanized or mechanically fastened to the belt surface. The construction of the conveyor is conventional on the carrying side, with the trough angle limited by the stiffness of the belt and the configuration of the cleats. The return idlers must account for the cleats through the use of split rolls, rubber disk rollers, or wing rollers. In some cases belts have low profile chevrons–1/2 inch (12 mm) or less–attached to or indented into the surface. Generally these belts do not require special return idlers.

The primary advantage of the cleated belt is its ability to increase the conveyor's angle of incline to as high as 45 degrees while preventing the roll back of the bulk material.

The principal disadvantage of a cleated conveyor is that as the angle of incline increases, the capacity

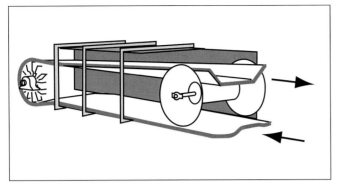

Figure 1.7
Enclosed roller conveyors incorporate ribs to pull material to the discharge.

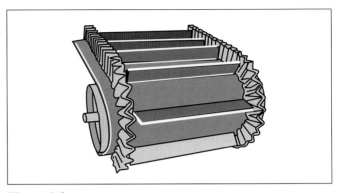

Figure 1.8
Pocket conveyors incorporate central ribs and flexible sidewalls to contain material.

decreases. The cleats are susceptible to damage, and they make the belt difficult to clean and seal. Special cleaners such as water sprays, air knives, beaters, or specially configured fingers are needed to clean the belt. For skirtboard seals to be effective, the cleats must stop short of the edge of the belt, and the belt must track well if skirt seals are to be used.

Typical applications for cleated belts are in aggregate and wood products. Cleated belts can be used as an alternative to steep angle conveyors when the capacity requirements are minimal. With some bulk solids, such as iron ore pellets, cleated belts are necessary even when using conventional incline angles.

Enclosed Roller Conveyors

These systems are totally enclosed conveyors where the belt is fitted with cleats or ribs on the carrying side. In addition to moving the load, these ribs are used to drag spillage and dust along the decking under the return run to the load zone. (**Figure 1.7**) There are various methods of self-loading the spillage and dust back onto the belt, usually with paddles attached to the tail pulley. Some designs have the idlers totally enclosed, while others have idlers either cantilevered from the outside or accessible from the outside.

The main benefit of these roller conveyors is the totally enclosed construction of both the carrying and return sides, which makes dust and spillage control easier. Other benefits include modular construction and lack of skirtboard maintenance.

Among the disadvantages of enclosed roller conveyors is that the dragging system for return of the dust and spillage is not effective when the bulk material is sticky. In some designs, maintenance of the idlers is difficult, and in all designs speed is limited by the mechanism that returns the spillage and dust to the load zone.

The most common application of roller conveyors is in the grain industry.

Horizontal Curved Conveyors

Conventional belt conveyors can be made to turn in a horizontal curve by elevating the idlers on the outside of the radius. Other than accommodating the elevated structure, no other major changes are necessary, and the conveyors use commonly available components. An engineering analysis is required to specify the correct belt, elevation, and tension for a given situation. The radius that can be turned varies but is commonly on the order of 100 yards (95 m).

The primary advantage of a horizontal curved conveyor is its ability to go around obstructions or geological formations. It may eliminate the need for a transfer point where a redirection of travel direction is required. The disadvantages of horizontal curve are the same as a conventional conveyor, and therefore, are minimal.

The best application of this technology is allowing long overland conveyors to adapt to the terrain and to eliminate transfer points. Design and construction has been developed to the point where it is common to see overland conveyors up to several miles (kilometers) in length incorporate multiple horizontal curves.

Pocket Conveyors

Pocket conveyors are similar to cleated conveyors in that they have large center cleats. (**Figure 1.8**) Flexible sidewalls are added to the belt, forming a continuous series of pockets similar to a bucket elevator. The belt is of special construction with a high cross stiffness to accommodate the necessary bend and return pulleys that are limited only on the outer edges of the belt. The pocket belt conveyor is often configured in an "S" shape and used in situations where there is limited space available.

There are several advantages to the pocket belt for low-to medium-capacity applications. They can lift the cargo vertically with minimal belt support. The belt can be twisted about the vertical axis to allow offset

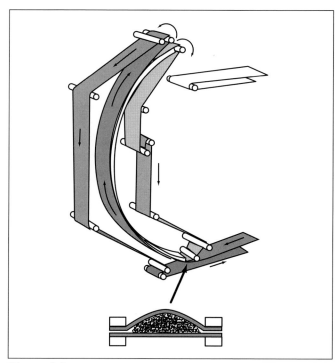

Figure 1.9
Steep angle conveyors sandwich the cargo between two belts.

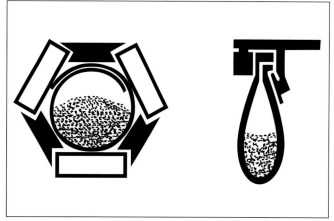

Figure 1.10
Tube conveyors form an enclosure around the material.

discharge configurations. Relatively tight convex and concave bends can be used. No side seals are needed.

The main disadvantage of pocket conveyors is that the belt is relatively expensive and must be custom-fabricated. This makes it advisable to keep a spare belt in inventory–which increases plant overhead. The system is most useful with small, dry lumps or dry, fine bulk solids. Like cleated belts, the pockets are susceptible to damage, and belt cleaning is difficult. Carryback will present significant problems with sticky materials.

Steep Angle Conveyors

Steep angle conveyors are constructed using two belts. The cargo is placed between the two belts like the meat in a sandwich. **(Figure 1.9)** These systems use modified idlers to capture the cargo and form the sandwich. They are often in an "S" shape configuration and are typically used to lift or lower loads vertically.

The advantages of steep angle conveyors are that they can transport high capacities up steep inclines, and so they are most often used where there is limited ground space or a drastic changes in elevation. They use many conventional conveyor components and provide cost-effective solutions in some applications.

The disadvantage of these systems is that they place additional idlers and belt-guiding equipment in a small space, which can make maintenance difficult. Steep angle conveyors are sensitive to proper loading to prevent the cargo from falling back down the belt. The maximum angle that the cargo can be transported is determined by the characteristics of the cargo and the belt. Steep angle conveyors present special belt cleaning challenges, as they must be effectively cleaned to maintain capacity, and the top belt must be cleaned in an "upside-down" position. Care must be taken in the design stage to allow adequate space for installation of belt cleaners and for the return of the carryback into the system, especially from the top or "cover" belt.

Typical applications for steep angle conveyors are for in-pit crushers where they are a cost-effective means of elevating the material when compared to longer-distance overland conveyors or haul trucks. Another typical application is for filling silos when the lack of ground space or the existing conveyor layout makes it impossible to use a conventional conveyor. Steep angle conveyors are also used extensively in self-unloading ships as the method of bringing the cargo from the hold to the discharge conveyor.

Tube Conveyors

This class of specialty conveyors uses a specially designed belt and carrying system. **(Figure 1.10)** In one configuration the belt has special edges that can be captured by an overhead carrying system similar to a trolley conveyor. Another design features a specially constructed belt that is formed into a tube by a series of radially-placed idlers. In both cases the belt is formed into a sealed, dust-tight, tube-like shape. The belt is opened up at the loading and discharge points with special guides.

Tube conveyors can adapt to many site conditions with the ability to incline as much as 30 degrees and turn tight horizontal and vertical curves. They are ideal for tight spaces and for bulk solids that must be kept contamination free. They can handle most bulk solids at reasonable flow rates.

The disadvantages of these systems are the added cost of the structure and guides and the need for specially designed belting. There are some difficulties in sealing the loading points, but the discharge points are conventional. Tube conveyors need an adequate cleaning system to keep the carrying system hardware from fouling and to maintain the capacity of the conveyor.

Typical applications are for port sites where contamination and spillage are major concerns or in industrial applications where space is limited. Materials that are commonly carried on tube conveyors include coal and chemicals.

The Standard for Mechanical Conveying

These alternative conveying systems may be suitable for particular installations or solve one particular problem. However, each poses other limitations and drawbacks. For general purposes, the conventional troughed idler belt conveyor is the performance standard and the value leader against which the other systems must be evaluated. Troughed idler belt conveyors have a long history of satisfactory performance in challenging conditions.

The overall success of a belt conveyor system greatly depends on the success of its transfer points. If material is loaded improperly, the conveyor will suffer damage to the belt, to rolling components, and/or to its structure that will decrease operating efficiency. If material is allowed to escape, it will cause numerous maintenance headaches, again leading to reduced production efficiency and increased operating and maintenance costs.

The Challenge of the Transfer Point

A typical transfer point is composed of metal chutes that guide the flow of material. **(Figure 1.11)** It may also include systems to regulate the flow, to properly place the flow within the receptacle (whether belt, vessel, or other equipment), and to prevent the release of fugitive material.

Transfer points are typically installed on conveyors to:
- move the material to or from storage or process equipment.
- change the horizontal direction of the conveyor system.
- divert the flow to intermediate storage.
- allow effective drive power over a distance that is too long for a single conveyor.

The method and equipment for loading the belt contribute much toward prolonging the life of the belt, reducing spillage, and keeping the belt trained. The design of chutes and other loading equipment is influenced by conditions such as the capacity, size, and characteristics of material handled, speed and inclination of the belt, and whether it is loaded at one or several places. An ideal transfer loading point would be designed to load the material on the belt:
- centrally.
- at a uniform rate.
- in the direction of belt travel.
- at the same speed as the belt is moving.
- after the belt is fully troughed.
- with minimum impact.

At the same time, it would provide adequate space and systems for:
- edge sealing and back sealing.
- carryback removal.
- dust management.
- inspection and service activities.

But In Real Life....

But these goals are difficult to achieve. The accommodations required by real life are likely to lead to problems. This is because transfer points have a much greater effect on, and value to, the operation of the conveyor than just that of a hollow vessel through which material is funneled. The loading point of any conveyor is nearly always the single most critical factor in the life of the belt. It is here where the conveyor belt receives most of its abrasion and practically all of its impact. It is also here where the forces that lead to spillage or dust creation act on the material and the belt.

The problem is that transfer points are a center for the interaction of many and conflicting forces, some arising from the material passing through, and some from the belts that run into and out of them. Material characteristics, air movements, and impact levels add forces that must be addressed by any system designed to halt the escape of fugitive material. In addition, many requirements imposed by the plant's overall operating environment will subject transfer points to additional forces and limitations.

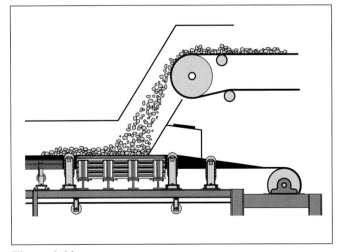

Figure 1.11

At transfer points, conveyors are loaded or unloaded.

The Engineering of Transfer Points

There are three basic approaches taken to the design of transfer points. The first and most common is the conventional method of drafting a solution to fit the master layout of the conveyors. This would be the *drafted solution*. The second method is to specify the critical components of the transfer point and design the overall conveyor layout to minimize transfer point problems. This is the *specified solution*.

The third method is to analyze the characteristics of the bulk solid and engineer custom chute work, which minimizes the disruption of the bulk solid trajectory and places the material on the next belt in the proper direction and at the speed of the receiving belt. This is the *engineered solution*. This third class of transfer point is typified by specifications that require the bulk solid to be tested for its flow properties. The transfer of material from one belt to another is engineered using fluid mechanics to minimize the dust and spillage. Similar re-engineering can be done on existing transfer points. **(Figure 1.12)**

Specifications for an engineered transfer point should include:
- material characteristics and flow rates.
- minimum performance requirements in terms of hours of cleanup labor and/or amount of spillage per hours of operation.
- maximum budget requirements for annual maintenance and periodic rebuild at supplier-specified intervals.
- ergonomic requirements of access for cleanup and maintenance.
- engineering drawings and specifications for wear parts and complete maintenance manuals.

One Step Forward, One Step Back...

Unfortunately, improving the operation of complex equipment like a conveyor transfer point is not a question of solving one narrowly defined problem. Rather, the attempt to solve one problem in a sophisticated operating system typically uncovers or creates another problem. This second problem can prove as difficult to solve as the original, if not more so.

It is never easy to solve multiple-cause, mutiple-effect problems with fugitive materials and achieve total material control. Any one component can at best solve one aspect of a complex problem. One step–if taken alone–may provide some incidental or even substantial improvement in transfer point performance. However, its performance will probably not reach the level specified nor the longevity required. Other aspects of the problem will remain and by their existence downgrade the performance of the new individual component.

Figure 1.12
This transfer point has been re-engineered to improve performance.

For example, a new edge-sealing system may provide an immediate improvement in the prevention of material spillage from a transfer point. However, if there is no inner-chute wear liner present, the force of the weight of material on the skirting will create side pressure that abuses the new seals, leading to abrasion and premature failure. Eventually, the amount of spillage returns to its previous unacceptable level. The operations, engineering, and maintenance departments will list the newly installed sealing system as a failure; "another product that did not match up to its salesman's promises." And worse, spillage will continue to extract its high price from the efficiency of the conveyor and the overall operation. The same is true for dust suppression and collection equipment installed without proper material containment systems.

The Systems Approach

The key to any engineering improvement is a detailed solution that encompasses all components of the problem. The costs of undertaking such a systems approach will prove higher than those of upgrading any single component. But the return-on-investment will justify the expense.

Talking systems engineering is easy; it is the application of this comprehensive approach that proves difficult. Development of a comprehensive approach requires the incorporation of values, including knowledge of material and process, the commitment of resources, and consistent maintenance to keep the system operating at peak efficiency and achieve total material control.

Belting for Conveyors

While a belt conveyor system is composed of a number of components, none is more important than the belt itself. The belt represents a substantial portion of the cost of the conveyor, and its operation may be the key factor in the overall (economic) success of the entire plant. Therefore, the construction of the belt must be selected with care, and all measures must be employed to safeguard the integrity of the belt.

This section focuses on what is called "black belt." This refers to the heavy-duty belting typically found in bulk material handling. Even though it is available in other colors, it is described as "black belt" to distinguish it from the belting used in packaged goods and other light-duty applications.

Conveyor belting is composed of two parts: the carcass and the covers.

The Belt Carcass

The carcass is the most important structural section of the belt, as it must contain the tensile strength to handle the material load placed on the belt. The primary purpose of the carcass is to carry the tension necessary to move the loaded belt and to absorb the impact energy unleashed by the material as it is loaded onto the belt. No matter what belt support system is employed, if the belt carcass cannot handle the initial impact energy, the belt will prematurely fail. The carcass must be adequate to allow proper splicing techniques and be strong enough to handle the forces that occur in starting and moving the loaded belt. The carcass also provides the stability necessary for proper alignment and support idlers.

Most carcasses are made of one or more plies of woven fabric, although heavy-duty belting now incorporates parallel steel cables as reinforcement. Carcass fabric is usually made of yarns woven in a specific pattern. The yarns that run lengthwise (parallel to the conveyor) are referred to as warp yarns and are the tension-bearing members. The transverse or cross yarns are called weft yarns and are primarily designed for impact resistance and general fabric stability.

Years ago, conveyor belts typically used yarns made of cotton as the textile reinforcement. For improved cover adhesion and abuse resistance, a breaker fabric was often placed between the cover and the carcass. Through the 1960s and 1970s, carcass reinforcements underwent a change until today, most belts are made with fabrics of nylon, polyester, or combinations of the two. These fabrics are superior to the older fabrics in nearly all respects, including strength, adhesion, abuse resistance, fastener holding, and flex life. Breakers are rarely needed or used with these belts, because little or no improvement is achieved in either adhesion or abuse resistance. Presently, aramid fabrics are gaining in conveyor belt usage and offer high strength, low elongation, and heat resistance.

Carcass Types

There are four types of belt carcasses:

1. *Multiple-ply belt carcasses:* This type is usually made up of two or more plies, or layers, of woven belt fabric, are bonded together by an elastomer compound. Belt strength and load support characteristics vary according to the number of plies and the fabric used.

   The multiple-ply conveyor belt was the most widely used through the mid-1960s, but today it has been supplanted by reduced-ply belting.

2. *Reduced-ply belting:* These belts consist of carcasses with either fewer plies than comparable multiple-ply belts or special weaves. In most cases, the reduced-ply belt depends upon the use of high-strength synthetic textile fibers concentrated in a carcass of fewer plies to provide higher unit strength than in a comparable multiple-ply belt.

   The technical data available from belt manufacturers generally indicates that reduced-ply belting can be used for the full range of applications specified for multiple-ply belting.

3. *Steel-cable belting:* Steel-cable conveyor belts are made with a single layer of parallel steel cables completely imbedded in rubber as the tension element. The carcass of steel-cable belting is available in two types of construction: all-gum construction uses only cables and cable rubber, while the fabric-reinforced construction has one or more plies of fabric above and below the cables but separated from the cables by the cable rubber. Both types have appropriate top and bottom covers.

   Steel-cable belting is produced using a broad range of cable diameters and spacing, depending primarily on the desired belt strength. This type of belting is often used in applications requiring operating tensions beyond the range of fabric belts. Another application is in installations where take-up travel limitations are such that changes in the length of a fabric belt due to stretching cannot be accommodated.

4. *Solid-woven belts:* This type of belting consists of a single ply of solid-woven fabric, usually impregnated and covered with PVC with relatively thin top and bottom covers. Abrasion resistance is provided by the combination of PVC and the surface yarns of the fabric. Some belting is produced with heavier covers and, thus, is not dependent on the fabric for abrasion resistance.

Top and Bottom Covers

Covers protect the carcass of the belt from load abrasion and any other local conditions that contribute to belt deterioration. In a few cases, these conditions may be so moderate that no protection and no belt covers are required. In others, abrasion and cutting may be so severe that top covers as thick as 1/2 inch (12 mm) or more are required. In any case, the goal of cover selection is to provide sufficient cover to protect the carcass to the practical limit of carcass life.

The covers can be made of a number of elastomers, from natural and synthetic rubbers, PVC, or materials specially formulated to meet application requirements with specialized features such as oil resistance, fire resistance, or abrasion resistance.

The top and bottom covers of the conveyor belt provide very little, if any, structural strength to the belt. Their purpose is to protect the carcass from impact damage and wear and to provide a friction surface for driving the belt. Usually, the top cover is thicker than the bottom cover due to its increased potential and more durable for abrasion, impact damage, and wear.

The pulley side cover is generally lighter in gauge than the carrying side because of the difference in wear resistance needed. The difference cannot be too great or the belt will cup. Some belts, however, have the same gauge of cover on each side.

Users sometimes turn the belt over when one side –the carrying side– has become worn. In general, it is better to avoid inverting the belt after deep wear on the top side, because this inversion presents an irregular surface to the pulley, which results in poor lateral distribution of tension. In addition, there may be material fines embedded into the belt's former carrying surface, so when the belt is turned over, this material is now placed in abrasive contact with pulley lagging, idlers, and other belt support systems.

Grades of Belting

Various national and international bodies have established rating systems for the belting used in general-purpose bulk material handling. Designed to guide end users as to what grades to use in different applications, the ratings specify different laboratory test criteria without guarantee of performance in a specific application.

In the United States, the Rubber Manufacturers Association (RMA) has established two standard grades of belting covers.

RMA Grade I belting is designed to provide high cut, gouge, and abrasion resistance. These covers are recommended for service involving sharp and abrasive materials and severe impact loading conditions. RMA Grade II belting covers are designed to provide good to excellent abrasion resistance but not as high a degree of cut and gouge resistance as Grade I belting.

The International Standards Organization has similarly established a grading system under ISO 10247. This standard includes Category H (Severe Cut and Gouge Service), Category D (Severe Abrasion Service), and Category L (Moderate Service). Category H is roughly comparable with RMA Grade I; Categories D and L approximate RMA Grade II belting.

In addition, there are belting types constructed to meet the requirements of stressful applications such as service with hot materials, service in underground mines, or service with exposure to oil or chemicals. There are new energy-efficient covers. These covers reduce the tension required to operate the belt because there is less roller indentation resistance as the belt moves over the idlers. Belting manufacturers should be consulted to determine which belting type is most appropriate for each application.

Cut Edge or Molded Edge

There are two types of edges available for belting: molded edges or cut edges. Steel cable belting is manufactured to a pre-determined width, and so has molded edges. Fabric ply belting is available with either a molded edge or a cut edge.

A molded edge belt is manufactured to the exact width specified for the belt, and its edges are enclosed in rubber. Accordingly, the carcass fabrics are not exposed to the elements. Because a molded edge belt is made for a specific order, it will probably require a longer lead-time and is typically more costly than a cut edge belt.

A cut edge belt is manufactured and then cut or slit down to the specified size to fulfill the order. That way, the manufacturer may hope to fill two or three customer orders out of one piece of belting produced. This makes cut edge belting more cost effective (and hence economical) to manufacture, so this type of belt has become more common. The slitting to the ordered width may occur at the time of manufacture, or it may be done when a belt is cut from a larger roll in a secondary operation–either at the manufacturer or at a belting distributor.

A cut edge belt can be cut down from any larger width of belting. This makes it more readily available. However, there are some drawbacks. At the cut edge(s), the carcass of the belt is exposed and is therefore more susceptible to problems arising from environmental conditions in storage, handling, and use. In addition, the slitting process is vulnerable to problems such as dull slitting knives, which can lead to problems such as belt camber. In addition, there are the unknowns that come with buying used or re-slit belting.

Specifying a Belt

The selection and engineering of the proper belting is best left up to an expert, who might be found working for belt manufacturers, belting distributors, or as an independent consultant. A properly specified and manufactured belt will give optimum performance and life at the lowest cost. Improper selection or substitution can have a catastrophic consequence.

There are a number of operating parameters and material conditions that should be detailed when specifying a conveyor belt, including:
- Limit variations in thickness to +/- 5 percent.
- Limit camber to +/- 1 percent of belt width in 100 meters.
- Belt surface to be smooth, flat, and uniform +/- 5 hardness points.
- Hours of operation loaded and unloaded.
- Details of the transfer point including trough angle and transition distance as well as information on material trajectory and speed.
- Description of material to be handled as completely as possible, including lump sizes and proportion and material temperature range.
- Description of belt cleaning system to be used.
- Description of chemical treatments (i.e. de-icing agents/dust suppressants) to be used.
- Description of atmospheric contaminants (from nearby processes or other sources).
- Specification of local weather extremes that the belt must withstand.

In the specifications, the user should require that the manufacturer's mark be molded into the bottom rather than the top cover, where it will not interfere with belt cleaning and sealing systems.

Compatibility with Structure

Buying belting is like buying a man's suit. To fit you best, it must be tailored to your construction. Conveyor belts are designed for different lengths, different widths, different trough angles, and different tensions. A belt must be compatible with the conveyor structure, and there is more to compatibility than belt width.

But unfortunately, this is not commonly understood at the operating level. Too often, there is a "belt is belt" philosophy in place. This originates from an incomplete understanding of the complexity of the belting problem. This philosophy becomes practice at times when there is a need to economize or to provide a faster return to service. The response in these cases is to use belting out of stock, either a leftover piece or a spare belt found in maintenance stores that the company wants to get out of inventory. Or it can be a belt from an outside source, like a belting distributor or a used equipment dealer.

It is a false economy to use a "bargain" belt that is not fully compatible with the conveyor system. Incompatibility of belt to structure is a common problem leading to poor belt performance and a poor return on the belting investment. It may be the most common problem found on conveyors where a replacement belt has been installed or pieces have been added to the existing belting.

Understanding the basics of compatibility is essential to ensure good performance of the belt and conveyor.

Belt Tension Rating

A belt's rated tension is a measurement of the force required to overcome the friction of conveyor components, to overcome the weight of the load, and to raise the load through any elevation on the conveyor. Each belt is rated as to its strength (the amount of pulling force that it will withstand.) The strength of a belt (or more accurately, the tension it is able to withstand) is rated in Pounds Per Inch of Width, commonly abbreviated as PIW. The metric equivalent of PIW is Newtons per Millimeter (N/mm).

This rating is a function of the reinforcement included in the carcass of the belt, the number of and type of material in the fabric plies, or if it is a steel cable belt, the size of the cables. The belt's top and bottom covers provide very little of the belt's strength or tension rating.

For instance, a 3-ply belt may have each ply rated at 110 PIW. This would then translate to a belt with a PIW rating of 330. The rated tension (in pounds) that the belt can be operated without damage would be 330 multiplied by the belt width (in inches). Putting more pressure–in the form of material load, take-up weight, and incline gravity–against this belt would cause severe problems, up to breakage.

The higher the rated tension of the belt, the more critical the compatibility of the belt and structure becomes.

Figure 2.1

A cupped belt will not lie flat on the conveyor's return run.

Minimum Bend Radius

Bending a belt over a radius that is too small can damage the belt. This may result in separation of plies or cracking of the belt's top cover. Inadequate pulley size can also lead to the pullout of mechanical splices. Belting is designed with a minimum pulley size specified by the manufacturer. This is determined by the number and material of plies, whether they are steel or fabric reinforced, the rated tension of the belt, and the thickness of the top and bottom cover.

At the original development of a conveyor system, the desire to use a thicker belt on a conveyor system (to extend belt life in the face of high load zone impact, for example) may lead to the installation of larger pulleys.

A common mistake occurs when an operation notices some type of surface damage to the carrying side of the belt. The immediate reaction is to install a thicker belt on the conveyor in the expectation of getting a longer service life. But this thicker belt itself will be damaged as it tries to wrap around pulleys that are too small. If the thicker belt has a minimum pulley size that is larger than the pulleys on the structure, the belt may actually yield a shorter life–the problem it was installed to solve.

Trough Angle

Belts are troughed to allow the conveyor to carry more material. As the trough angle is increased, more material can be carried.

All belts have the ability to be formed into a trough by idlers. However, all belts also have a maximum angle of trough that can be formed without doing permanent belt damage. Again, the type of belt carcass, the thickness of the belt, the width of the belt, and the tension rating of the belt determine the angle of trough.

Troughability is shown as a minimum belt width allowed for the various trough angles.

Exceeding the maximum trough angle of a particular belt can cause the belt to permanently deform into a cupped position. **(Figure 2.1)** This cupping can make a belt difficult to seal, difficult to clean, and almost impossible to track. As the belt cupping increases, the surface contact between the rolling components and the belt is reduced, diminishing the ability of the rolling component to steer the belt in the desired direction. This will also reduce the friction between the drive pulley and the belt, which means the conveyor may not operate at design speed.

Problems that may occur if the belt's troughing capability is exceeded are damage to the top and bottom covers and damage to the carcass in the idler junction area. The belt may not lie in the trough correctly, creating sealing and tracking problems if the belt trough angle is exceeded.

If a belt is too stiff and will not properly trough, it will not steer (track) properly through the system. This will be quickly seen as spillage off the sides of the conveyor and damage to the edges of the belt.

Transition Distance

Typically, the belt travels across the tail pulley in a flat position. As the belt leaves the tail pulley and moves into the loading zone, the belt edges are elevated, forming the trough where the material is carried. This is done with transition idlers–that is, idlers set at intermediate angles between flat and the conveyor's final trough angle.

(There is a similar–but reverse–transition area at the conveyor's head pulley, where the conveyor is taken from troughed to flat profile just before it reaches the discharge point).

As the belt is formed into a trough, the outer edges of the belt are stretched more than the center of the belt to travel around the idlers. If the transition is formed too quickly, damage in the idler junction area of the belt may occur.

Factors affecting the required transition distance are belt width, the type of belt, and the rated operating tension of the belt.

Although there are some typical 'rules of thumb' for transition distance, the belting manufacturer should be contacted to ensure the correct distance is used.

When installing a new belt on an existing structure, it is important to check the conveyor's existing transition distances. It is common to see that a conveyor has a transition area that is shorter than what is ideal. So it is even more critical not to increase the problem by applying a replacement belt that requires a longer transition distance.

It might be possible to lengthen a conveyor's transition area. There are two ways to do this. One is moving the tail pulley back to extend the distance before the load zone. The second method is adapting a two-stage transition area, where the belt is partially troughed before it enters the loading zone and then completes its transition to the final trough angle after the load has been introduced. *For more information on transition areas, see Chapter 4: Tail Pulleys and Transition Areas.*

More commonly, however, circumstances–like the available space and the available budget–preclude this. The most economical and suitable option is to make sure that the belting is suitable for the existing transition area.

Know Your Structure, Know Your Belt

Placing any belt on a structure without understanding the characteristics of the belt and the structure is asking for trouble. A conveyor belt must be compatible with the system it is installed on. Just placing a belt on a structure without understanding the important characteristics of the belt will impair the performance of the system and reduce the life of the belt. There can be problems in the form of mistracking, shortened belt life, broken and pulled-out splices, downtime, and added maintenance expense.

There are two opportunities to match a belt to a conveyor. When the structure is being built, the belting desired can be used as a determinant in the selection of pulley sizes and trough angles. After the conveyor structure is completed, the characteristics of the existing structure should be the determining factor in the selection of belting to be run on that conveyor.

It is recommended that all parameters of the belt and structure be fully understood prior to selecting and installing a belt on an existing structure. A complete review of the conveyor may be required to ensure that the belt used on the system is the right choice.

Figure 2.2
Improper storage practices can lead to damaged belting and poor performance on the conveyor.

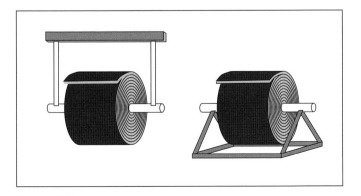

Figure 2.3
Supporting the roll of belting with a cradle or stand will prevent uneven stress.

Storage and Handling of Conveyor Belting

The conveyor belt has long been the most economical and most efficient form of bulk material handling for many industries. However, if this important part of the of plant is to perform as expected, it must be carefully stored and handled from the time of its manufacture until the time of its installation on the conveyor system. Improper storage techniques, like storing the belt on its side and/or on the ground, can lead to a damaged belt that will perform poorly when on the conveyor structure. **(Figure 2.2)** As the length of time stored increases, and as the size of the roll of belting increases, so does the importance of the following correct procedures.

The costs of handling, shipping, and storing the conveyor belt are minor compared to the purchase price of the belt, and, therefore, the correct procedures should be followed regardless of the size of the roll or the length of time stored.

Rolled on a Core

As the belt leaves the manufacturer or the supplier, the belt should be rolled carrying side out on a solid core. The core gives the belt protection from rolling it in a diameter that is too small, and protection when the belt is lifted through the center. It also provides a means for unrolling the belt onto the conveyor. The core size is determined by the manufacturer, based on the type, width, and length of the roll of belt. The core size can be smaller than the minimum pulley diameter, as the rolled belt is not in a tensioned state.

The core should have a square hole through it for insertion of a lifting bar.

Properly Supported

The conveyor belt should never be stored on the ground. Ground storage concentrates all the weight of the roll onto the bottom. The belt carcass is compressed in one small area and is usually not compressed equally from side to side. The carcass may be stretched more on one side or the other. This may be a possible cause of belt camber. (Belt camber is a banana-like curvature of the belt, running the length of the belt).

Under no circumstances should a roll of belting be stored on its side. The weight of the roll may cause that side of the belt to expand, creating camber problems. Moisture may migrate into the carcass through the cut edge of the belt, creating carcass problems or belt camber.

The belt should be supported upright, on a stand, off the ground. This places the stress of the roll's weight on the core, relieving some of the load on the bottom. **(Figure 2.3)**

This support stand can be utilized during shipping to better distribute the weight of the belt. The support stand can then be utilized in the plant for storage, or the belt can be transferred to an in plant storage system that properly supports the roll of belt.

Rotated

If the support stand is designed correctly, the roll of belt can then be randomly rotated every 90 days. This will more evenly distribute the load throughout the carcass. The reel of belting should be marked at the factory with an arrow to indicate the direction of rotation. Rotating the belt in the opposite direction will cause the roll to loosen and telescope.

Properly Protected

During shipping and storage, the roll of belt should also be covered with a tarp or dark plastic. Covering the roll of belt protects the belt from rain, sunlight, or ozone. The covering should remain in place during the entire storage process.

The roll of belt should be stored inside a building to protect it from the environment. The storage area should not contain large transformers or high voltage lines that may affect the belt. The building does not have to be heated, but it should be somewhat weather tight and insulated from extreme conditions.

Properly Lifted

When lifting a roll of belt, a square lifting bar of the correct size should be placed through the core. Slings or chains of the correct size for the weight of the roll should be used. A spreader bar should be utilized to prevent the chains or slings from damaging the edges of the conveyor belt. **(Figure 2.4)**

The Importance of Belt Conservation

As noted above, the cost of the belt will easily exceed the cost of other conveyor components and may, in fact, reach the level where it matches the cost of the steel conveyor structure.

Obviously, all systems installed around the conveyor–whether to feed it, receive material from it, or assist in

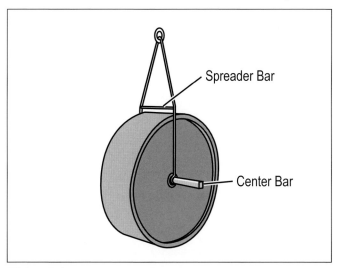

Figure 2.4
The use of a spreader bar when lifting the belt will prevent damage to belt edges.

its material transport function–must be designed to present the minimum risk to the belt.

A key to providing a reasonable return on the investment in belting is avoiding damage and prolonging its service life.

Belt Damage: Causes and Cures

Damage to belting can be a major drain on the profitability of operations using conveyors. This expense–which occurs regularly in plants around the world at costs of thousands of dollars–can often be prevented. Unfortunately, relatively little effort is put into the analysis of the life of the belt and the reasons for a belt not reaching its optimum life.

The types of belt damage can be divided into two groups: Normal Wear and Avoidable Injury. In addition, there are a number of types of damage such as weather exposure and chemical attack that are outside the scope of these discussions.

Wear due to the normal operation of the conveyor can be managed and minimized to prolong the belt life, but a certain amount of wear is considered acceptable. Perhaps avoidable damage cannot be totally prevented, but it can be minimized through proper design and maintenance management.

The first step in preventing belt damage is to identify the source(s) of the damage. A step-by-step analysis can almost always lead to the "culprit."

Types Of Belt Damage

Here is a brief review of the major types of belt damage.

Impact Damage

Impact damage, as its name implies, is caused by large, sharp conveyed material striking the top cover of the belt. (**Figure 2.5**) The result of this impact is a random nicking, scratching, and gouging of the top cover. A large frozen lump of coal may cause this type of damage. If severe enough, the belt can actually be torn completely through. This type of damage is usually seen under crushers or in mines on conveyors handling ROM (Run of Mine) material.

Long material drops without some method to help the belt absorb the energy can also lead to impact damage. *The methods to reduce loading zone impact damage are discussed in Chapter 5: Chutes and Chapter 6 Belt Support.*

Entrapment Damage

Entrapment damage is usually seen as two grooves, one on each side of the belt, near the edge of the belt where the belt runs under the steel conveyor skirtboard. (**Figure 2.6**)

Many times this damage is blamed on the pressure from a skirtboard sealing system. However, extensive study has shown this type of belt damage is more likely

Figure 2.5
High levels of impact and sharp material edges can put nicks in the cover of the belt.

Figure 2.6
Material can become wedged into the steel skirtboard and abrade the belt.

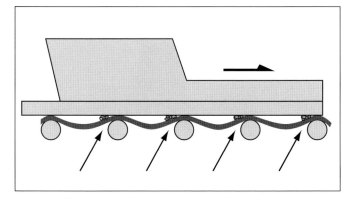

Figure 2.7
Belt sag will create material entrapment points.

due to the conveyed material becoming entrapped between the sealing system and the conveyor belt.

This material entrapment occurs when the belt is allowed to sag below the normal belt line and away from the sealing system. Material becomes wedged into this "pinch point," forming a spearhead to gouge or abrade the surface of the belt as it moves past. **(Figure 2.7)** This leads to any of several negative events:

- The trapped material will form a high-pressure area, causing excessive wear on the sealing system (seen as scalloping in the seal at each idler). **(Figure 2.8)**
- Grooves will be worn along the entire length of the belt under the skirtboard.
- Material will be forced off the sides of the belt, leading to piles of material spillage under the load zone.

Material entrapment can also be caused when the skirting is placed inside the chute wall in the path of the material flow. Not only does this type of arrangement cause material entrapment and belt damage, it also reduces the cross-sectional area of the chute wall, in turn, reducing conveyor capacity.

The only way to prevent belt sag is to use bar support systems to support the belt and stabilize the path in the entire skirted area. *For additional information, see Chapter 6: Belt Support.*

Belt Edge Damage

Edge damage is usually seen as frayed edges on one or both sides of the belt. **(Figure 2.9)** If edge damage is not identified and corrected, it can be severe enough to actually reduce the width of the belt to a point where it will no longer carry the rated capacity of the conveyor.

Belt mistracking is probably the leading cause of belt edge damage. There are numerous reasons why a belt might mistrack. These causes range from out-of-alignment conveyor structures or off-center belt loading or accumulations of material on rolling components, or even the effect of the sun on one side of the belt.

There are many techniques and technologies that can be used to train the belt. These would include laser surveying of the structure, adjustment of idlers to counter the belt's tendency to mistrack, and installation of self-adjusting belt training idlers that use the force of the belt movement to steer the belt's path.

The key to good belt tracking is to find the cause for the mistracking and then remedy that cause, rather than spending time and money turning one idler one direction and another idler a different direction in pursuit of better tracking. *For more on controlling belt wander, see Chapter 15: Tracking.*

Worn Top Cover

This type of damage is seen when the top cover of the belt is worn in the load carrying area of the belt or even across the entire top. Several factors can contribute to worn top covers.

Carryback is one cause. This is material that clings to the return side of the conveyor belt and randomly drops off along the conveyor return. If not controlled, this fugitive material can build up on the ground, confined spaces, and rolling components. These accumulations can quickly build to a point where the belt runs through the fugitive material, and the cover will wear away. This damage will happen more quickly on materials with sharp-edged particles and higher abrasion levels.

Faulty belt cleaner selection and improper cleaner mounting can also lead to top cover damage. Belt cleaners must be mounted properly to avoid chattering. Belt cleaner chatter can quickly remove the top cover of the belt if not corrected immediately.

Research has shown that even properly installed belt cleaning systems can cause some wear on the cover of the belt. This would qualify as "Normal Wear" of the belt. With properly tensioned cleaning devices, this wear is modest and has been shown to be less than the abrasion from one idler seizing due to material buildup.

Slow moving, feeder-type belts that convey materials from vessels under high "head loads" can also suffer top cover damage. Reducing the head forces on the belt will reduce the potential for damage.

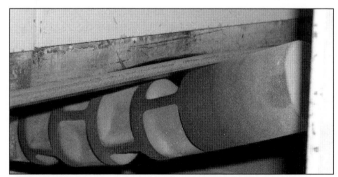

Figure 2.8
High pressure areas above idlers can cause a scallop of wear in the sealing system.

Figure 2.9
Damage to the belt edge is a sign of the belt mistracking into the conveyor structure.

Rips and Grooves from Foreign Objects

This type of damage is caused by stray pieces of metal, ranging from packing crate strappings to the teeth from loader buckets. These metal pieces can become wedged into the conveyor structure, forming a knife to gouge or slit the belt. This damage can be the most difficult type to control, because it occurs very quickly and often with catastrophic effects. There are a number of ways to minimize but not totally eliminate the amount of "tramp iron" in the material flow. These methods include grizzly screens, metal detectors, and video monitors. Regardless of the effectiveness of the precautions, the belt is still vulnerable.

Top Cover Cracking

Short random cracks in the top cover running perpendicular to the belt travel may be caused by a mismatch between the belt and the supporting structure.

Each belt, depending on the manufacturer, number of ply, reinforcing materials, and thickness, will require a different minimum bend radius. This type of damage occurs if the belt is not matched to all pulley diameters of the structure. Bending a belt in too small of a radius will cause stress on the top cover. This undue stress on the top cover will cause the rubber to crack, exposing the reinforcing materials of the belt, and also lead to damage to the belt's internal carcass.

Any change from the original belt specifications should also be done in conjunction with a study of the conveyor's pulley diameters.

Installing a thicker belt on an existing system to improve life by preventing other types of belt damage, such as impact, may dramatically shorten the life of the new belt if the diameters of the conveyor pulleys are smaller than recommended by the manufacturer. Always check with the belt manufacturer to ensure this design parameter is met.

Heat Damage

Conveying hot materials also may cause the top cover to crack or plies to separate. The cracks from heat damage may run either parallel or perpendicular (or both) to the direction of belt travel. If the conveyed material is hotter than the belt can handle, holes may be burned through the belt. Using high-temperature belts may reduce this heat cracking and increase belt life. The only true solution is to cool the material prior to belt conveying or to use some other method of conveying.

Junction Joint Failure

As the belt travels the conveying system, going from flat to troughed at the terminal pulleys, the outer edge of the belt is required to travel farther on the outer one-third of the belt than the center one-third of the belt.

Figure 2.10

Junction joint failure is caused by an improper transition distance.

Figure 2.11

A cupped belt will not lie down on the idlers.

Figure 2.12

Viewed from the top, camber is a longitudinal curve in the belting.

Thus the outer one-third of the belt must stretch more than the inner one-third. If this stretching takes place in too short of a distance, the belt can become damaged where the outer wing rolls meet the center flat roll. This damage is termed junction joint failure.

Junction joint failure is seen as small stretch marks running the entire length of the conveyor in the area where the wing roller and the flat rollers of the idlers meet. **(Figure 2.10)** These stretch marks run parallel with the belt, approximately one-third of the belt width in from each edge. This type of belt damage may be so severe as to actually tear the belt into three separate pieces.

Junction joint failure is caused by too short of a transition distance wherever the belt goes from trough to flat or flat to trough. Belt thickness, reinforcing materials, materials of construction, and trough angles all determine the transition distance. When designing a new system, contemplating a change in belt specifications, or increasing trough angles, check with the belt manufacturer to ensure the proper transition distances are maintained at both the head and tail pulleys.

Belt Cupping

Belt cupping happens when the belt has a permanent curvature across its face, perpendicular to the line of travel. **(Figure 2.11)** Belt cupping can be caused by heat, by transition distances not matched to the belt, or by too severe of trough angle for the type of belt being used. Over-tensioning the belt can also cause it to cup.

Belting manufacturers specify a maximum trough angle for each belt. Chemicals can also cause a belt to cup up or down, depending on whether the chemical extracts or swells the elastomer in the belt's top cover.

Cupped belts are extremely difficult to track, as the frictional surface between the belt and the rollers is drastically reduced.

Belt Camber

Belt camber is a longitudinal curve of the belt if viewed from its top side. **(Figure 2.12)** Camber is often confused with a crooked splice. Camber produces a slow side to side movement of the belt, while a crooked splice causes the belt to jump quickly from side to side.

This type of damage can be created during manufacture or from improper storing, splicing, or tensioning of the belt. Proper storage and handling are essential from the time of manufacture to the time of installation.

A crooked splice has a short area of influence, while camber is from one end of a belt section to the other. If a belt is composed of more than one section, it may have more than one camber.

Repair of Conveyor Belting

For most operations, conveyor belt life is measured in years. To achieve the lowest operating cost, inspection of the belt should be a scheduled maintenance procedure. Any belt damage noted during these inspections should be repaired promptly to prevent small problems from becoming big trouble.

Types of Repairable Damage

Inspection will note a number of different types of belt damage, each with its own distinct cause. **(Figure 2.13)** The major types of repairable damage include:

Grooves
> This superficial damage is caused by abrasion from bulk solids onto the belt's top cover. It will show itself as worn patches or "bald spots" on the cover, which will become weaker and weaker, allowing the entry of water or contamination into the belt carcass.

Longitudinal Rips
> These are caused by a fixed object—like a metal bar that is jammed into the conveyor structure. This object will slit the belt like a knife, and this slit can rapidly extend the full length of the belt.

Profile Rips
> These are seen at the belt edge as small tears that move inwards in the form of a rip. They are caused by the belt rubbing against a hard surface, i.e., the belt mistracking into the structure, usually limited to short incisions from the edge inwards.

Edge Gouges
> These are caused by blunt objects pulling or tearing chunks of rubber out of the belt at the

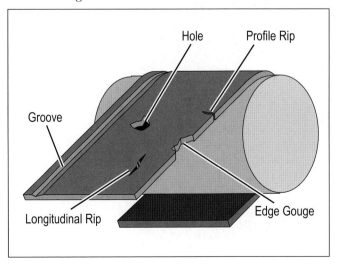

Figure 2.13
Types of Conveyor Belt Damage.

edge; again, typically caused by mistracking into the structure. The exposed edge of the belt is weaker than the top, and the carcass fibers can typically be seen, especially in a cut edge belt.

Damage to belting permits the entrance of moisture or foreign materials, and thus promotes premature failure of the belt. To preserve the belt, it is important to make prompt and effective repairs of any damage.

Vulcanized repairs can be made during scheduled maintenance outages when sufficient conveyor downtime is available to allow the long time required to make a vulcanized joint. But in nearly all cases, a vulcanized repair requires removal of a complete section of belt, and then either re-splicing the remainder or adding a new saddle of belting.

Fortunately, most forms of damage lend themselves to relatively simple methods of repair. Such repairs can be made with self-curing, adhesive-like repair materials to keep moisture or foreign material out of the carcass. Mechanical fasteners are another method for repairing damaged belting to restore service without significant downtime and replacement of expensive belting, and to extend the service life of expensive belts.

To preserve the belt by minimizing damage, it may be wise to invest in a rip detection system.

Belt Repair Using Adhesives

Adhesives provide a cost-effective means to repair conveyor belting with a high quality bond. Use of adhesive compounds will save downtime and money in maintenance budgets without requiring heavy vulcanizing equipment (or creating obstructions with hardware in the belt).

Adhesive repair compounds offer cost-effective solutions for belt maintenance that are durable, reliable, and easy to use. There are a number of products available to do this. They include solvent-based contact cements, heat-activated thermoplastics, and two-component urethane elastomers.

Figure 2.14
Two-component urethane belt repair adhesives are mixed and then spread on the belt like frosting.

All of these systems require some degree of surface preparation, ranging from a simple solvent wipe to extensive grinding or sandblasting. They may need a separate application of a primer to improve adhesion.

Most commonly used for standard cold-vulcanized splices, solvent cements are also used for bonding repair strips and patches over damaged areas.

Thermoplastic compounds are "hot melts" that are heated to a liquid state, and then harden as they cool to form a repair. As they cool quickly from their application temperature of 250 to 300° F (120 to 150° C), the repair must be performed quickly before the adhesive returns to the hardened (non-adhesive) condition. Problems encountered with thermoplastic adhesives include the possibility of shrinkage as the adhesive cools and high-temperature conditions that may cause softening of the adhesive, that leads to joint failure.

Urethane products are two-component systems that the user can mix and then spread like frosting directly onto the area to be repaired. **(Figure 2.14)** They typically achieve operating strength in a short period–1 to 2 hours–but will continue to cure for 8 to 12 hours until full strength is reached.

All adhesive systems offer fairly simple applications, assuming the instructions are followed. Of course it is critical that the adhesive manufacturer's instructions be

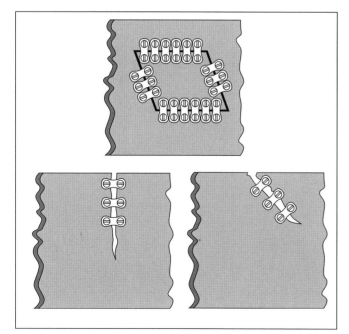

Figure 2.15
Damaged belts can be repaired using mechanical fasteners.

followed carefully as to surface preparation, component mixing, pot life, application technique, and cure time. The length of time for operating and full cure may be the requirement that should be used as the standard for selecting any particular product.

It is important to get the profile of the repaired area down to match the profile of the original belt in order to avoid more damage to the belt (and the "popping out" of the repair).

And it is also important to resolve the cause of the problem–remove the obstruction or correct the mistracking–which led to the belt damage in the first place. Otherwise, the resumption of operations after repair merely initiates a waiting period until the damage recurs, and the repair must be made again.

Mechanical Fasteners for Belt Repair

Because of their comparative ease of installation, mechanical splices are often used in emergency repair situations when a new piece must be added to an old belt or when a belt must be patched or a rip closed. **(Figure 2.15)** In these cases, the mechanical fasteners are used as a "band-aid" to cover damage and close a hole, allowing the conveyor to begin running again.

Repair is a good use for mechanical splices, providing care is used to properly install and recess the fasteners. It must be remembered, however, that these repairs are only temporary stopgaps and are not designed for permanence. It is always important to solve the root cause of the problem in order to prevent recurrence.

Recovery from belt damage does not have to involve lengthy downtime. Using a mechanical rip repair fastener method can speed up the return to service and reduce the cost. Mechanical rip repair fasteners offer an inexpensive and fast (often 20 minutes or less) repair. They can be installed with simple tools and without discarding any belting. As soon as the fasteners are in place, the belt can be returned to service without waiting for any extended "cure" time. They can be installed from the top side of the belt without removing the belt from the conveyor.

Splice suppliers recommend use of alternating two- and three-bolt fasteners for repair of jagged rips. **(Figure 2.16)** Put the two-bolt side of the three-bolt fastener on the "weak" flap side of the rip to provide greater strength. For straight rips, standard two-bolt mechanical fasteners are acceptable.

One-piece, hammer-on "claw" style fasteners **(Figure 2.17)** can provide temporary rip repair where the speed of repair and return to operation is critical.

Rip repair fasteners can also be used to fortify gouges and soft, damaged spots in the belting to prevent these spots from becoming rips.

Preserving the Life of the Belt

Considering the investment in conveyor belting, the importance of preserving the belt through regular inspection and repair activities cannot be overstated. The relatively minor costs for belt repair, and the somewhat more significant expense of an outage to allow any needed repair, will be paid back many times over by the savings provided by an extended belt life.

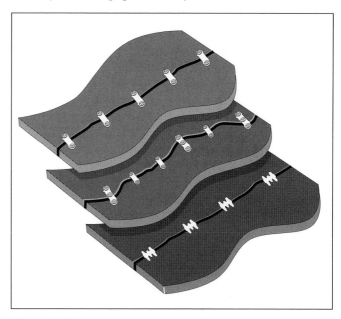

Figure 2.16
Jagged rips can be repaired with rip repair fasteners installed in a pattern that alternates two- and three-bolt fasteners.

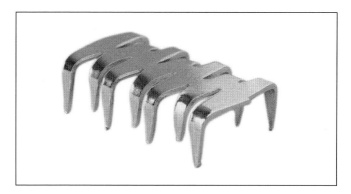

Figure 2.17
Claw-type fasteners can be hammered in place for a fast repair of a ripped belt.

chapter 3

Splicing the Belt

Conveyor belting is shipped on a roll, and so before use, the two ends of a belt must be joined together–spliced–to provide a continuous loop. The two methods for joining the ends of the belt together are vulcanizing and mechanical splices.

Vulcanized Splices

Vulcanizing is generally the preferred method of belt splicing, due to the superior strength and longer service life it offers. However, operations that necessitate frequent additions or removals of belting, such as underground extensible belts or conveyors with limited take-up capabilities (that require pieces of the belt be removed to maintain proper belt tension), are not suitable for vulcanized splices.

With their superior strength, vulcanized splices also allow the application of maximum belt tension, resulting in better belt-to-pulley traction for efficient operation. Vulcanizing also provides a cleaner conveyor, because it does not allow material fines to filter through the splice as can happen with mechanical joining systems. A properly-performed vulcanized splice does not interfere with rubber skirting, idler rolls, belt support structures, or belt cleaners.

There are two types of vulcanizing–hot and cold–currently in use. In hot vulcanizing, the layers of a belt are stripped in a stair-step fashion and overlapped with glue and rubber. A "cooker" or heated press is then used to apply heat and pressure to "vulcanize" the belt to form the endless loop. In cold vulcanization (also called chemical bonding) the belt's layers are joined with an adhesive or bonding agent that cures at room temperature.

A vulcanized splice has no internal weaving, braiding, sewing, welding, or other mechanical link. The tensile members of the splice–textile plies or steel cords–do not touch each other. The splice is solely dependent on adhesion at the internal interface. Although designs differ according to belting and application, a vulcanized splice basically consists of putting a material between the two pieces of belt and additional material over the top of the joint. Adhesion is often obtained through an intermediary rubber or rubber-like material called tie gum, installation gum, foil, or cement.

Typically, the materials used for a vulcanized splice–consisting of cement, tie gum, noodles, or cover stock, all depending on belt style and construction–are available in kit form. Kits from the belt manufacturer are sometimes preferred, although there are generic kits available for the most common belt grades. The materials in the kit are perishable; they have a shelf life in storage and also a "pot life" when they are mixed into the ready-to-apply state.

> "Improper application of a splice will shorten the life of the belt and interfere with the conveyor's operating schedule."

Splice Geometry

There are a number of different splice designs used today. In general, the geometry of a splice can be the same whether the splice will be hot- or cold-vulcanized.

Depending on the belt material and conditions of service, a splice can be installed with a "finger" joint at a bias angle or across the belt perpendicular to the belt.

If the splice is laid across the belt at an angle, this bias angle is generally 22 degrees. This angle increases the length of the bonding surface and reduces stress on the splice as it wraps around the conveyor's pulleys.

It is important that the overlap and any added materials are properly installed to prevent belt cleaning systems from damaging the splice by tearing out the filler strips.

Steps in Vulcanizing a Belt

The actual step-by-step procedures for vulcanized splices vary between manufacturers (**Figure 3.1**). But in general, there are three steps.

1. *Preparation of the Belt Ends*
   In this step the belt ends are cut at the correct angle and then stripped or pulled apart to expose the various plies to be joined.
2. *Application of Cement, Gum, or Other Intermediary Material*
   This step provides the building up–much like the making of a sandwich–of the materials that will form the completed splice.
3. *Curing of the Splice*
   The assembled materials are pressed together and cured, through time or through the application of heat and pressure over time, to form the finished splice.

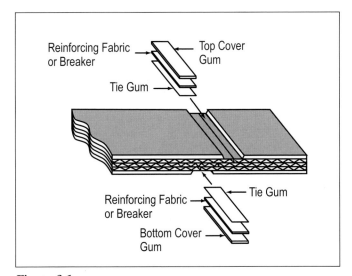

Figure 3.1

A vulcanized splice is made from a sandwich of belt and added materials.

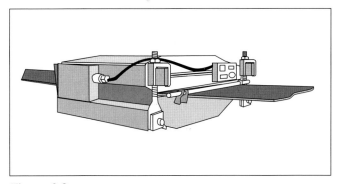

Figure 3.2

A press applies heat and pressure to form a vulcanized splice.

Hot Vulcanization

In hot vulcanization, a special press (**Figure 3.2**) applies both heat and pressure to the splice to cure the intermediary and cover materials into a high-strength joint. Pressure can range from 5 psi to 175 psi (34 to 1200 kPa), depending on belt and heat ranges from 280° to 400° F (120° to 200°C), depending on belt type and rubber compound.

Portable vulcanizing presses for curing the splice are available in a range of sizes to match various belt widths. Small fabric belt splices can often be cured in a single heat, but larger splices are commonly cured in two, three, or more settings of the vulcanizer without problems. With steel cord belts, however, it is important that the press be large enough to cure the splice in a single heat to avoid undesirable rubber flow and cord displacement.

The time required for cure will depend on belt thickness and compound. Belt manufacturers should include time and temperature tables in splicing manuals. Although the equipment is automated, the process may require constant human attention to achieve best results.

Again, the belting manufacturer is the best source of information on proper splicing technique and materials.

Cold Vulcanization

Cold vulcanization is also called chemical bonding. Here, the belt is joined using adhesives or bonding agents that will fuse the ends of the belt together. In general, cold vulcanization can be applied on any belt that is step, butt, or overlap spliced.

In cold splicing, splices are not cured in a press. The belt ends are laid carefully together in proper alignment and full contact achieved with hand roller, pressure rollers, or hammering in a prescribed pattern. The bond can often be improved by simply putting weights on the belt during the cure interval. Cure time is as fast as one hour, although better results can be achieved if the belt can be left for at least six hours.

The cold vulcanization splice technique offers some advantages over hot vulcanization. There is no heating source or press required, the equipment is easier to transport, and no special electricity is required. Therefore, cold vulcanized splices can be performed even at remote sites where access is difficult and no special electricity can be provided.

Only small hand tools are required, and so the costs for purchase and maintenance of the installation equipment are low.

The downtime required by the cold vulcanizing process is much less than needed for the hot vulcanization, which will save money in applications with continuous operating requirements.

Disadvantages of Vulcanized Splices

The disadvantages of vulcanizing that must be considered are the higher initial cost and the length of time required to perform the splice.

Generally speaking, a vulcanized splice takes longer to install than a mechanical splice, therefore, requiring a longer shutdown period. The peeling back of layers of belting to prepare for vulcanization can be difficult. It can take up to 12 hours to return to service by the time the belt is prepared and the press heated and cooled sufficiently to allow the belt to be handled.

As vulcanization often involves an outside contractor, the cost is higher than work performed by in-house maintenance personnel. This longer completion time will be particularly troubling (and expensive) in cases where an emergency repair is required to allow the resumption of operations. In this case, the delay required to secure an outside crew and equipment increases the cost by extending the outage and adding "emergency service" surcharges.

Vulcanizing can be more difficult and less reliable on older, worn belts.

In applications on belts moving hot materials, the belt should run empty before stopping. A hot load left on a stopped belt can "bake" a splice and reduce its life.

Installation of a vulcanized splice can consume a considerable length of belt–eight to ten feet (2.4 to 3 meters), in some cases–particularly when using a bias splice. This may require that a longer belt or a new section of belt–typically called a saddle–be added. (And of course, inserting a saddle means adding yet another splice).

When designing new conveyor systems that will incorporate vulcanized belts, it is wise to include a generous take-up length. This will accommodate belt stretch and maintain proper tension while avoiding need to shorten the belt with a time-consuming new splice.

Even though a vulcanized splice is more expensive and time-consuming to perform, it is an excellent investment. Splicing work done by reputable firms is of high quality workmanship and materials and is typically guaranteed. For permanent splicing of the belt, a vulcanized splice is often the best choice.

Mechanical Fasteners

On the market today there are many types of mechanical fasteners available for belt splicing. These work on the principle of pinching the two ends of the belt together between metal fasteners with a staple, hinge and pin, or solid plate design. Mechanical fasteners are now fabricated from a variety of metals to resist corrosion and wear and match application conditions.

For many years, mechanical splices were considered "the ugly stepchild" or the low-quality alternative to vulcanization as a method of joining the belt. But recent developments have moved mechanical fasteners into a position that is virtually equal with vulcanization. These innovations include the use of thinner belts (made possible by the use of synthetic materials in belting), improvements in fastener design and fastener materials to increase strength and reduce wear, and the development of tools to recess the profile of the splice.

Advantages of Mechanical Splices

The principal advantage of mechanical splicing is that it allows the belt to be separated to allow extension or shortening of the belt in applications like mining; or to allow service to other conveyor components, such as pulley lagging, idlers, or impact cradles.

Another advantage of mechanical fasteners is that they minimize repair downtime by installing in an hour or two, while vulcanized joints can take half a day or more. Fasteners are easily installed by available plant maintenance personnel, using only hand tools or simple portable machines; whereas, vulcanizing usually requires calling in independent contractors with specialized equipment. The fastened joint will cost a few hundred dollars and consume only a few millimeters of belting; whereas, the vulcanized joint can cost several thousand dollars and consume several meters of belting.

The working strength of the fastened joint is primarily limited by the belt's mechanical fastener rating (its ability to hold onto a mechanical fastener), which is expressed in pounds per inch width (PIW) or kilonewtons per meter (kN/m). These ratings vary depending on the type of belt construction, but typically are calculated at around one-tenth of the belt's ultimate breaking strength for a conservative margin of safety. Consequently, it is important to make sure that fastener selection observes the manufacturer's recommendations for belt mechanical fastener ratings.

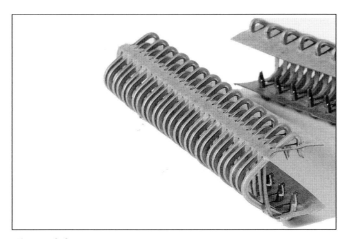

Figure 3.3

Wire hook fasteners are generally considered a light-duty splice.

Types of Mechanical Splices

Mechanical fasteners are available in three classes–wire hook fasteners, hinged fasteners, and plate fasteners–with fastening options within each group.

Wire Hook Fasteners

Wire hook belt fasteners–the original steel lacing–replaced rawhide as the belt joining system for continuous belts in the days when belts were more commonly used for power transmission than for conveying. **(Figure 3.3)** These systems are still commonly seen on lightweight belting in applications from grocery store checkout belts to airport baggage handling systems.

These wire hook fasteners have been generally disqualified from bulk solids operations because they are less resistant to abrasion, lack the compression advantages of plate types, and allow more sifting of material through the splice. However, newer designs are available using wire hook or common bar lacing design that can be applied up to at least the medium range of bulk material applications.

Hinged Fasteners

A second category is hinged fasteners. Here, each segment consists of a pair of top and bottom plates connected at one end by their hinge loops. **(Figure 3.4)** This keeps the plates in correct alignment above and below the belt. The plates are drawn together with the securing of the splice fasteners. Hinged fasteners are available in three methods of attachment: staple, bolt, or rivet.

Hinged fasteners are usually supplied in continuous strips to fit standard belt widths. These strip assemblies ensure proper spacing and alignment. The strips are fabricated so pieces can be snapped off to allow the remainder to fit non-standard belt widths.

The chief advantage of hinged fasteners is that by removing the linking pin they are separable. This way the belt can be shortened, extended, removed, or

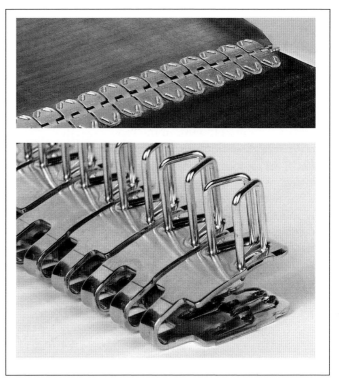

Figure 3.4

(Top) Hinged fasteners are supplied in continuous strips to match belt width.

(Bottom) To insert or remove belt sections, hinged fasteners allow the belt to be separated.

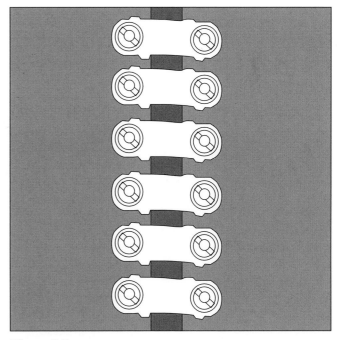

Figure 3.5

Solid plate fasteners provide a sturdy joint for heavy-duty application.

opened to allow maintenance on conveyor components.

Hinged fasteners provide several other benefits. Installation on the two belt ends can be done separately and even performed off site. Belts of different thicknesses can be joined together using fastener halves matched to their respective belt half.

Solid Plate Fasteners

The final type of mechanical splice is performed with solid plate fasteners. **(Figure 3.5)** This class of fastener makes a strong, durable joint with no gap for fines. Solid plate fasteners are effective in the most rugged conveying applications in mines, quarries, and steel mills. In applications where the belt is thicker than 7/8 inch (22-mm), solid plates are the only choice. However, solid plate fasteners are intended for permanent joints only and not recommended for belts in applications that require opening of the belt joints to change belt length or location.

Solid plate fasteners are typically provided in loose individual pieces, in boxes, or buckets. The plate segments are installed from one belt end to the other using staples, rivets, or bolts.

Bolt-fastened solid plate fasteners have some unique advantages. They can be applied diagonally across the belt to allow use on pulleys that are smaller than recommended size. They can also be installed in a V-shaped pattern **(Figure 3.6)**, may be the only choice for joining thick, high-tension belts designed for vulcanization.

One problem with bolt-fastened solid plate fasteners is that they typically use only two bolts on each plate, with one on each side of the splice. Tightening down on the ends of the splice means the leading and trailing edges are more compressed than the middle of the plate. This allows the middle to crown, creating a wear point in the fastener and in belt cleaners or other systems that touch the belt.

Another problem arises if the conveyor uses pulleys that are smaller than 12 inches (300 mm) in diameter. In this case, solid plate fasteners may be too large to bend around the pulley, causing components of the splice to pull out or break.

Problems with Mechanical Splices

If the materials to be conveyed are hot, the transmission of heat through a metal fastener may be a factor that leads to the selection of a vulcanized splice. When the material temperature exceeds 250° F (121° C), the amount of heat passed through the metal fastener into the belt carcass can weaken the fibers, ultimately allowing the fastener to pull out. In these applications, a vulcanized splice should be preferred.

Failure to inspect fasteners and resulting failures may be a cause of severe belt damage. Pulling out of fasteners on a portion of the belt width can start longitudinal ripping of the belt. When belt and fasteners have been properly selected, pullout is usually due to insufficiently tight bolts or worn hooks or plates. Note that plate-type mechanical fasteners typically allow individual replacement of damaged plates, which if done when damage is first observed, often eliminates the need to cut out and replace the entire joint.

Using the wrong size or type of mechanical fastener can greatly reduce the operating tension capacity of the belt. The extra thickness of a mechanical splice that has not been recessed or that is of the wrong specification will make sealing the transfer point almost impossible. Splices that are oversized and too thick to pass through the transfer point area can catch on the wear liner or chutework, abusing the splice and shortening its life. Often the fasteners used in the splice will not be properly trimmed, and these extended rivets or bolts can catch on other components.

Most, if not all, mechanical splices will allow some small quantity of the conveyed material to filter through the joint itself. This material will fall along the run of the conveyor, resulting in cleanup problems and the potential for damage to idlers, pulleys, and other conveyor components.

Plate-type fasteners, in a well-made joint, are quite free of material leakage. The hook-type and other hinge-type fasteners are all subject to sifting through of fine material, which is eliminated with vulcanized splices.

Proper Installation

Mechanical splices can be installed relatively easily by plant personnel; but as a consequence, they can easily be misapplied, particularly by untrained personnel or in an emergency "get running in a hurry" situation. It is critical that plant personnel be trained in the proper installation of mechanical fasteners.

It is a common but incorrect practice to stock only one size of mechanical splice in the maintenance supply room. Over the years, the specifications for the belts used within a plant may have changed, but the

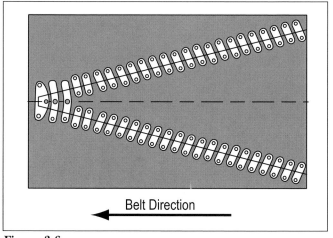

Figure 3.6
This V-shaped splice may be the best choice for joining high-tension belts with mechanical fasteners.

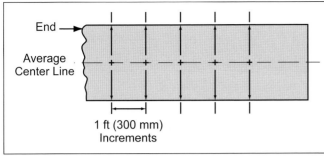

Figure 3.7

To determine the belt's average centerline, connect the center of five different measurements across the belt.

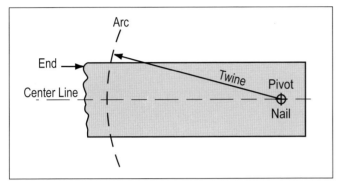

Figure 3.8

In the "double arc" method, first draw an arc from a pivot point two or three times the belt width from the belt end.

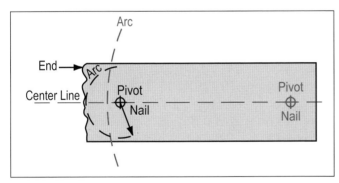

Figure 3.9

Then draw a second arc so that it intersects the first arc near the belt edge on both sides.

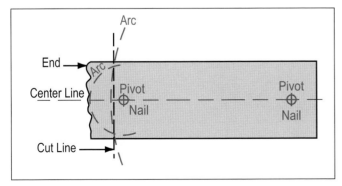

Figure 3.10

Finally, mark the cut line by connecting the points where the arcs intersect on both sides of the belt.

mechanical fasteners kept on hand in the storeroom stayed the same, which can lead to a variety of problems.

Selecting the Proper Fastener

Most fasteners are available in a range of sizes. In all cases, the manufacturer's recommendations should be checked to ensure that the fastener size is matched to the pulley sizes and belt thickness.

Remember, if the belt is to be skived in order to countersink the fastener down to the surface of the belt, this skived thickness should be considered when thinking about fastener size.

Make sure the diameter of the smallest pulley belt will wrap 90 degrees around to ensure compatibility.

Fasteners are available in a variety of different metals to meet the requirements of special applications. These properties include non sparking, nonmagnetic, abrasion-resistant, and corrosion-resistant materials. Hinge pins are available in similar selection. The manufacturer should be contacted for the proper recommendation to be used with any specific application.

Training for selection and installation of splices should be carried out by supplier personnel. When installed in accordance with manufacturer's instructions, a mechanical splice can provide a very economical method of joining the belt. When incorrectly specified or applied, mechanical splices can create expensive and recurrent problems.

Squaring the Belt End

Where belt ends are joined with mechanical fasteners, the first requirement of a good joint is that the belt ends be cut square. Failure to do so will cause some portion of the belt adjacent to the joint to run to one side at all points along the conveyor.

Using the belt edge as a squaring guide is not recommended, although new belts usually can be squared with sufficient accuracy by using a carpenter's square and working from the belt edge. Used belts may have an indistinct edge due to wear, and so the following procedure is recommended:

1. Establish the belt's average centerline by measuring from the edges at five points along the belt, each roughly one foot (300 mm) farther from the end of the belt, and mark a series of points at the center of the belt. Connect this series of points using a chalk line or rule to determine the average centerline. **(Figure 3.7)**
2. Using a square, draw a line across the belt perpendicular to the average centerline. This line can be used for the cut line.

For greater accuracy or on belts with worn edges, you can employ a "double intersecting arc" method. After establishing an average centerline as above, pick a point on the centerline two to three times the belt width from the belt end. Using a string with a nail on the centerline as a pivot point, draw an arc across the belt so the arc crosses the edge of the belt on both sides. **(Figure 3.8)** Now, create a second pivot point on the centerline much closer to belt end. Strike a second arc (this one facing the opposite direction) across the belt so the second arc crosses the first arc on both sides of the average centerline. **(Figure 3.9)** Draw a line from the intersection of the arcs on one side of the belt to the intersection of the arcs on the other side. **(Figure 3.10)** This new line is perpendicular to the centerline of the belt and becomes the cut or splice line.

To check the accuracy of the squared end, measure back a given distance (say 48 inches or one meter) on both sides of the belt. Then measure diagonally from these new points to the end of the cut on the opposite side of the belt, so you are measuring a diagonal. **(Figure 3.11)** These two diagonal measurements should be identical. If you mark these two diagonal lines, they should intersect on the belt's centerline.

Notching the Trailing Side

To protect the corners of the belt, it is often useful to notch or chamfer the corners of the belts at the splice **(Figure 3.12)** On one-direction belts, it is only necessary to notch the trailing belt end. The notch is cut in the belt from the first fastener on each end of the splice out to the belt edge at a 60° angle. The notch will help prevent the hang-up of the corners of the belt at the splice on the conveyor structure.

Getting Along with Belt Cleaners

Plate type fasteners sometimes conflict with aggressive belt cleaning system cleaners, especially where hardened metal blades are used. Many operators prefer to use nonmetallic (i.e., urethane) belt cleaner blades on belts with mechanical splices for fear of wearing or catching the splice and ripping it out.

Two developments help make mechanical fasteners more scraper-friendly. One is the development of new tools for skiving that will easily remove a uniform strip of cover material to leave a smooth, flat-bottomed trough with a rounded corner to prevent tearing. These are much faster and safer than earlier methods that used knives or grinders.

A second development is new designs of "fastener friendly" cleaners, offering special blade shapes, materials, and mounting methods that minimize impact problems with fastener plates. Recently introduced "scalloped" mechanical clips are designed to allow belt cleaner blades to "ramp up" and over the plates without damage to cleaner or splice.

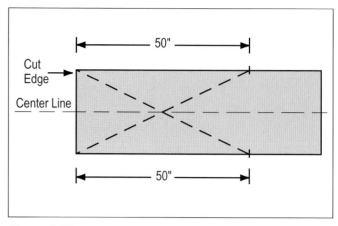

Figure 3.11

To check if the cut line is straight across the belt, measure diagonally from points the same distance on both sides of the belt.

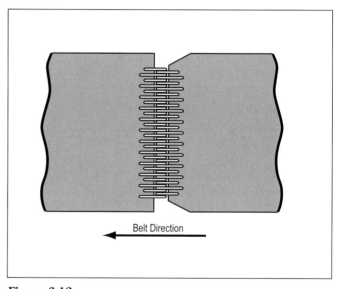

Figure 3.12

The belt's trailing edge should be cut away at the splice to prevent the corners catching on obstructions.

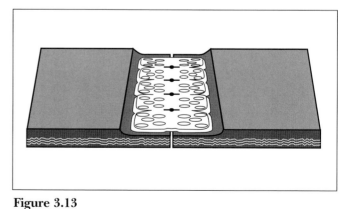

Figure 3.13

If properly recessed, both the top and bottom of the fasteners are below the surface of the belt.

There are no empirical studies on the wear of splices due to interaction with bulk material and with the cleaning and sealing systems. If good installation and maintenance practices are observed, the cleaning and sealing systems and the splices should be chosen on the basis of the performance required, rather than the worries over their life expectancy.

The Importance of Skiving

For a mechanical splice to function in a transfer point and allow effective sealing and cleaning, it must be properly installed. Proper installation means that both the top and bottom splice pieces should be recessed into the belt surface. **(Figure 3.13)**

Countersinking the fasteners–typically called skiving (rhymes with driving)–mounts the fasteners closer to the belt carcass fabric for a firmer grip and also makes the top of the fastener parallel with the belt cover to avoid belt cleaning systems.

Skiving the belt also reduces noise in operations, as clips are now recessed and do not "click clack" against the idlers.

This requires the top and bottom covers be cut down to the belt carcass. As the carcass provides the strength of the belt, and the top and bottom covers provide very little strength, this will not reduce the integrity of the belt or splice. Great care is required when skiving the belt, as any damage to the carcass of the belt can weaken the splice and therefore reduce the strength of the belt. When the splice is properly recessed, the metal components of the mechanical hinge will move without incident past potential obstructions such as impact bars, rubber edge skirting, and belt cleaner blades.

Dressing a Mechanical Splice

If for some reason–limited belt thickness, belt damage, or limited time to complete a repair–it is impossible to properly recess a mechanical splice, the splice can be dressed. This is the encapsulation of the splice in a material to protect both it and the cleaner from impact damage. **(Figure 3.14)** The cleaning system will still have to ride up and over the mechanical clips; the benefit is that the splice surface is smoother and there are no protruding obstacles in the path. The downside of this procedure is that because the mechanical splice is "buried" in the elastomer material, the splice is harder to inspect and repair.

Splice Inspection and Service

Where bolt-style fasteners are used, it is important that the plates be kept properly tightened. The most practical way to achieve this is to tighten the bolts so that the rubber behind the plate slightly swells. (Be careful not to overtighten fasteners or "bury" the plates in the belt cover, as this could cause damage to the plies of belting.) Manufacturers generally suggest retightening the fasteners after the first few hours of operation, and then again after the first few days of operation, and then at intervals of two or three months. This recommendation does not, however, preclude the practice of more frequent fastener inspection and more frequent bolt retightening if the inspection indicates this service is desirable.

The Importance of the Splice

Whether it is vulcanized or uses mechanical fasteners, a well-applied and maintained splice is critical to the success of the belt. Improper application of a splice will shorten the life of the belt and interfere with the conveyor's operating schedule. In the words of an old axiom: "If you can't find time to do it right, how are you going to find time to do it over?"

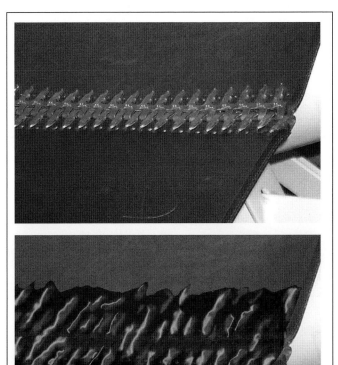

Figure 3.14

(Top) A non-recessed splice poses a risk to belt cleaners and other conveyor components.

(Bottom) Dressing the splice will protect both the mechanical fasteners and belt cleaners.

chapter 4

Tail Pulleys and Transition Areas

At the end of the conveyor's return run, the belt wraps around the tail pulley and moves up onto the top or carrying side. Here it must be prepared to receive the cargo before it enters the load zone. These preparations include stabilizing the belt path, centering the belt on the structure, shaping the belt into the desired profile for load carrying, and sealing the back and edges of the load zone to prevent spillage.

Care must be taken in accomplishing these tasks in order to minimize fugitive material now in the process (and later), as well as preserve the equipment and prepare the conveyor for maximum efficiency.

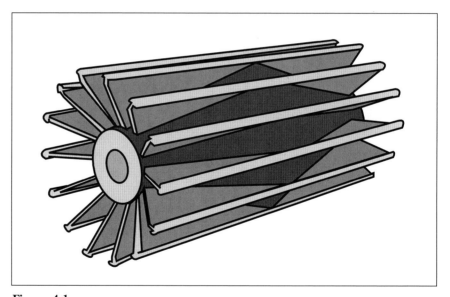

Figure 4.1
Wing-type tail pulleys prevent the capture of material between the pulley and the belt, but are hard on the belt.

"...it must be prepared to receive the cargo before it enters the load zone."

Centering the Belt

Having the belt in the center of the structure as it goes into the loading zone is a critical function. If the belt is not properly centered when it receives the load, the forces of loading will compound the mistracking and other problems encountered on the conveyor's carrying side.

If there is room between the tail pulley and the load zone, the belt can be centered by a tracking device installed in advance of the load zone. More commonly, however, the area between pulley and load zone is too short, so belt centering devices should be installed on the conveyor return to make sure the belt is centered as it enters (and exits) the tail pulley.

If the belt–centered at its entry to the pulley–mistracks between the pulley and the load zone, the problem is probably that the pulley is out of alignment.

Keeping the belt in the center is discussed in greater length in Chapter 15.

Tail Pulleys: Wings and Wraps

Wing-type tail pulleys are often installed as a method to reduce the risk of belt damage from the entrapment of lumps of material between the belt and the pulley. **(Figure 4.1)** Wing pulleys feature vanes that resemble the paddle wheel on a steamboat. This design allows material that would otherwise become trapped between a solid pulley and the belt to pass through the pulley face. Between the pulley's cross bars are inclined, valley-shaped recesses that prevent fine or granular material from being caught between the tail pulley and the return belt. These valleys provide a self-cleaning function–there is little surface area on which material can accumulate, and the turning of the pulley throws the material off the pulley's face. **(Figure 4.2)** If the conveyor is likely to spill some of its cargo onto the return belt, the wing pulley can act as an effective device for removing this spillage without belt damage.

Wing-type pulleys are also seen on gravity take-ups, where they offer the same benefits.

But despite their design intention, wing pulleys are still subject to buildup and entrapment and often do not provide the desired protection. They are most successful on slow-moving belts where cleaning and sealing are not critical requirements. Larger lumps of material can become wedged in the "wings" of the pulley, risking the damage the pulley was designed to avoid.

Wing-type tail pulleys also have a tendency to bend the belt quickly and can exceed the minimum bend radius of the belt. This may cause carcass damage to the conveyor belt.

The most significant drawback of wing pulleys is the oscillating action they introduce to the belt path. The wings introduce a pulsating motion that destabilizes the belt path and adversely affects the belt sealing system. It is counter productive to design a transfer point that emphasizes belt stability to minimize fugitive material, and then install a winged tail pulley that defeats all that effort.

A better choice than the conventional winged tail pulley is a spiral-wrapped tail pulley. **(Figure 4.3)** These pulleys have an additional steel strip wrapped in a spiral around the pulley circumference. The steel band is wrapped over the top of the wings in a spiral from the center out to each end of the pulley. Wrapping this band of steel around the wing pulley allows the pulley to provide the self-cleaning function, but eliminates the "bounce" imparted to the belt.

Figure 4.2
By vibrating the belt line, a wing pulley makes it difficult to seal the conveyor loading zone.

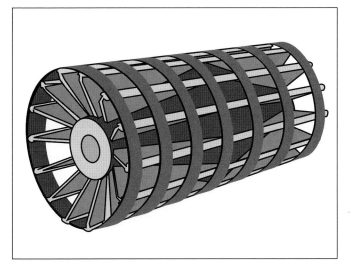

Figure 4.3
Wrapping a steel strip around the pulley will minimize vibration.

Figure 4.4
A spiral-wrapped tail pulley will reduce vibration in the belt line, while still preventing material entrapment.

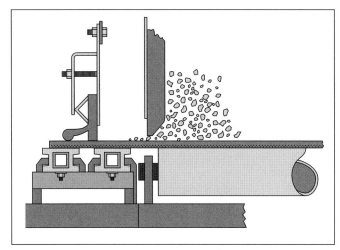

Figure 4.5
Flat belt feeder conveyors may incorporate inner and outer chute walls.

The wrap is applied in a spiral around the pulley, or perhaps two spirals, converging in the center from each end. **(Figure 4.4)** This design minimizes the potential for material entrapment and improves the throw of material.

Spiral wing pulleys are available as original equipment on new conveyor installations. Existing wing pulleys can be upgraded with a narrow, two-to-three-inch (50 to 75 mm) wide steel strip welded around the outside edge of the wings.

Flat Belts

Many bulk materials can be carried on flat belts. Flat belts are particularly common for materials with a steep angle of repose–the angle that a freely formed pile of material makes to the horizontal. With angles of repose above 30°, these materials range from irregular, granular, or lumpy materials like coal, stone, and ore, to sluggish materials (typically irregular, stringy, fibrous, and interlocking) such as wood chips and bark.

Flat belts are especially effective when the load (or a portion of the load) is to be discharged from the belt at intermediate points by plows or deflector plates.

Belt feeders use flat belts almost exclusively. This is because feeders are generally very short, and they must be fit into operations where there is little room to form the belt into a trough. Feeders typically operate with very high loads and use very heavy-duty idlers. Also, many feeder belts are reversing. To move the high material load, feeder belts often run at very high tension; it is difficult to trough a belt in this condition. In addition, this very high head load makes sealing difficult. By leaving generous edge distances and operating at slow speeds, the spillage can be controlled. In many cases, these belts are equipped with skirtboard

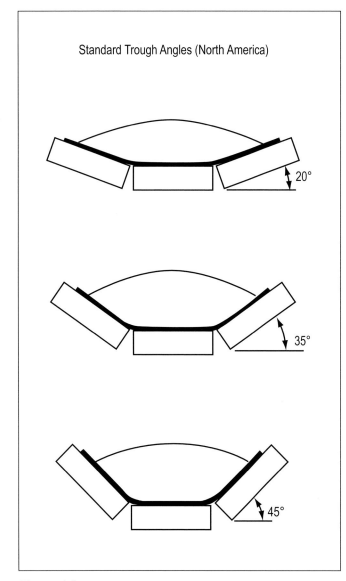

Figure 4.6
Belts are troughed to increase carrying capacity.

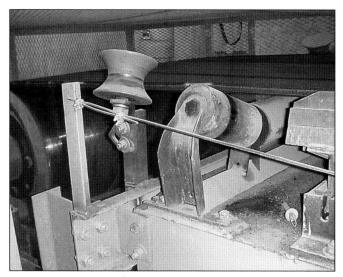

Figure 4.7
In the transition area, the belt profile is changed from flat to troughed.

and a sealing system along their full length. Other feeders use a dual chute wall design, where a space is left between the interior chute wall with wear liner and an outside chute wall that includes the belt's edge seal. **(Figure 4.5)**

Troughed Conveyors

Troughed belts can be used for carrying any bulk material suitable for a bulk conveyor. For most materials and most conveyors, the forming of the belt into a trough by bending the edges upward provides the significant benefit of a generous increase in the belt's carrying capacity.

This increase in capacity is seen in the conveyor outside the unconfined area of the transfer point. In the skirted area, however, the troughing of the belt (particularly to the higher angles) often reduces the conveyor's effective belt width and load capacity.

Troughing the belt typically makes a positive contribution to belt tracking.

Typical Trough Angles

The standard trough angles in Europe are 20°, 30°, and 40°, while in North America, trough angles of 20°, 35°, and 45° are common. **(Figure 4.6)** At one time the 20° trough was standard, but the deeper troughs have become more common as improvements in belt construction allow greater bending of belt edges without premature failure. In some special applications, such as high-tonnage mining, catenary idlers with a 60° trough are used to reduce spillage and impact damage.

Because of its lower amount of bend and the resulting reduction of belt stress, a 20° trough permits the use of the thickest belts, so the heaviest materials and largest lumps can be carried.

Troughing angles steeper than 20° are usually specified when the material has a low angle of repose. The 30° and 35° troughing angles are suitable for a very broad range of applications. They work best when allowances are made for constraints such as limitations in transition distance and the requirement for exposed edge distance for skirt sealing.

While they offer the benefit of greater capacity, the use of troughed belts also presents some limitations. These are areas that can present problems if not properly considered during the conveyor design and belting specification process. For example, most belts require a specified transition distance to prevent stress at the belt edges.

Transition

On a typical conveyor, the belt is troughed for the carrying portion of its journey and returned to a flat configuration for the return run. Consequently, at the terminal (head and tail) pulleys, the belt must be converted from a flat to a troughed position, or vice versa. This changing of the belt's profile is commonly called the transition. **(Figure 4.7)** Transitions exist at the tail (loading) and head (discharge) pulley locations of a troughed conveyor, and they can occur in other areas of the conveyor, such as a tripper head.

The length of a conveyor from the terminal pulley to the first fully-troughed idler is called the transition

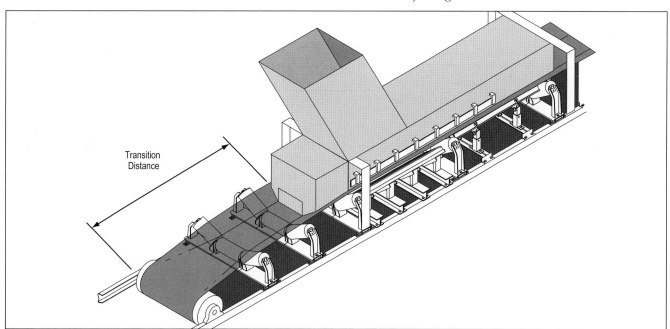

Figure 4.8
The area between the tail pulley and the first fully-troughed idler is called the transition area.

distance. **(Figure 4.8)** This area poses more potential risk to the belt than any other area of the conveyor. In changing from the flat belt to the fully troughed profile, tension at the sides of the belt is greater than at the center. This can cause the splice–either mechanical or vulcanized-fail at the belt edges. In addition, belt plies can separate.

The transition distance–the spacing allowed for this change in the belt's contour–must be sufficient at each terminal pulley. Otherwise, the belt will experience serious stress in the idler junctions (the points on a troughed idler set where the horizontal roller meets the inclined rollers). If the belt is forced into this junction area, the result may be a crease that may eventually tear along the entire length of belt. **(Figure 4.9)** Also, if the elasticity of the carcass is slightly exceeded, the belt may not tear but rather stretch beyond its limits, leading to belt tracking problems. If the transition distance is too short, an excessive difference between edge and center tensions can overcome the belt's lateral stiffness. This can force the belt down into the trough, so it buckles through the center, or catches in the idler junctions where the rollers of the idler join. **(Figure 4.10)**

These tensions can be kept within safe limits by maintaining the proper transition distance between the pulley and the first fully-troughed idler, thus, minimizing the stretch (and stress) induced in the belt.

To properly support the belt at these transitions, idlers with intermediate angles should be used between the terminal pulley and the first fully troughed idler. These transition idlers will allow the belt to gently change profile to the proper trough. Strain on the belt at the idler junction is then minimized, as it has been spread over several idlers.

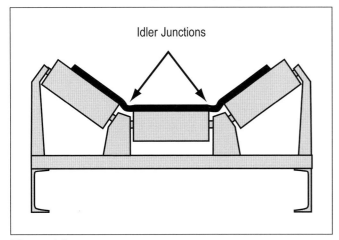

Figure 4.9
A transition distance that is too short can force the belt into the idler junctions, leading to damage.

Figure 4.10
Junction joint failure may require replacement of the entire conveyor belt.

Transition Distance

The distance required for this transition varies with the amount of troughing required, the belt thickness, the construction of the belt, and the rated tension of the belt.

The heavier the belt carcass, the more it will resist being placed in a troughed configuration. This is easy to understand if one remembers that a string stretched down the center of the conveyor will be shorter than the string placed on the outside edge of the idlers. The outer edges of the belt must travel farther than the middle of the belt. The higher the trough angle, the more the edges of the belt are stretched and the greater the distance required to reach that angle.

The required transition distance is a function of the construction of a belt. For the engineering of a new conveyor, the transition distance of the system can be designed to match belting previously selected to match the load and length characteristics of the conveyor. The belting manufacturer should be contacted to determine the specified transition distance

In the case of replacement belting for existing conveyors, the belt should be selected to match the transition distance provided in the conveyor installation. In no case should a belt be placed on a conveyor where the transition distance does not match the belt.

It is highly recommended that the supplier of the belt be contacted to ensure that the transition distance of the structure is compatible with the belt. Charts identifying the minimum transition distance as a function of the material trough depth and the rated belt tension for both fabric and steel cord belts at the various trough angles are published by CEMA in *Belt Conveyors for Bulk Materials*. However, the wisest course is to confirm the requirements with the belting supplier.

Transition Idlers

Depending on the distance, one or more transitional idlers should be used to support the belt between the terminal pulley and the first fully-troughed idler.

It is a good practice to install several transition idlers supporting the belt to work up gradually from a flat profile to a fully-troughed contour. Transition idlers can be manufactured at specific intermediate angles (between flat and fully troughed), or they may be adjustable to fit various positions. For example, it would be good practice to place a 20° troughing idler as a transition idler forward of a 35° troughing idler, and both a 20° and a 35° idler in front of a 45° idler.

It is also important to the stability of the belt (and the sealability of the transfer point) that the transition idler closest to the terminal pulley be installed, so the top of the pulley and the top of the idler's center roll are in the same plane. This is referred to as a full-trough conveyor.

The Problem with Half-Trough Pulleys

To shorten the required transition distance, the conveyor designer may be tempted to use a technique of raising the tail pulley. By elevating the pulley so its top in line with the midpoint of the wing roller (rather than in line with the top of the center roller), the required transition distance can be cut roughly in half. **(Figure 4.11)** This technique is usually employed to shorten the transition distance to avoid an obstruction or to save a small amount of conveyor length.

This half-trough technique has been accepted by belting manufacturers as a way to avoid excessive strain at the idler joint as a belt is troughed, particularly when fitting a conveyor into a limited space; but it can create problems. The half-trough design can cause the belt to lift off the idlers when it is traveling unloaded. As belt loading fluctuates, the belt line will change dramatically, so the transfer point cannot be effectively sealed. These shifts in the belt line create a "pumping action" that acts as a fan to push out airborne dust. In addition, this design can cause the belt to buckle in the transition area. Loading the belt when it is deformed in this manner makes effective sealing impossible and increases belt wear due to increased impact and abrasion on the "high spots" in the belt.

Solving the problems created by a half-trough belt design is more complicated than merely lowering the tail pulley to line up the center roll of the idlers. The transition distance must be maintained, so as the pulley

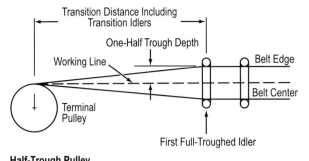

Half-Trough Pulley
Pulley is elevated, so its top is in line with the mid-point of the wing idler. This will shorten the transition distance, but allow the belt to rise off the idlers.

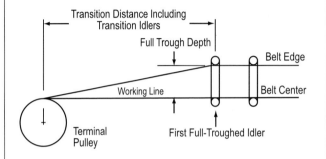

Full-Trough Pulley
Pulley top is in line with the top of the center roller of the troughed idler set. The transition distance will be longer, but the belt is more likely to remain down on the idler.

Figure 4.11

A half-trough technique is sometimes used to shorten the transition distance. This is not recommended.

moves down, it must also be moved further away from the pulley. If this is not possible, other possible changes include changing the trough angle in the loading area to shorten the required transition distance. The belt can then be changed to the higher trough outside the load zone. Another approach would be to adopt a very gradual transition area. Both of these techniques are discussed below.

Loading Areas

Loading while the belt is undergoing transition is bad practice and should be avoided. The area where the load is introduced to the belt should begin no sooner than the point where the belt is fully troughed. The belt should be properly supported by a slider bed or the midpoint of the first set of standard troughing idlers. It would be even better to introduce the load a foot (300 mm) or so beyond this fully troughed point in order to accommodate any bounce-back of material caused by turbulence.

If loading is performed while the belt is still in transition into the troughed configuration, the load is

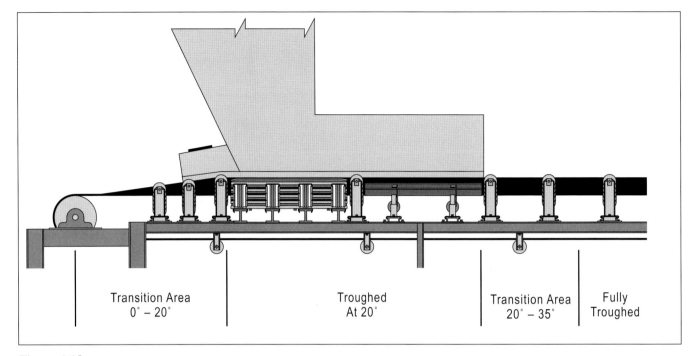

Figure 4.12
For conveyors where the loading point is too close to the tail pulley, two-stage troughing lets belt be loaded before it is fully troughed.

dropped onto a slightly larger area with non-parallel sides. This larger area increases pressure on the side skirts and increases wear of belts and liners as the belt forms its full trough. In addition, since the belt in the transition area is changing in contour, it does not have the stable belt profile required for effective sealing.

Material bouncing off other material and the walls of the chute can deflect behind the intended load point. Therefore, the trajectory of the material needs to be designed to contact the belt far enough up from the tail pulley to prevent material from flooding the transition area. The provision of adequate belt support in the loading zone ensures that the belt maintains a flat plane critical for effective sealing of the transfer point.

Two-Stage Transition Areas

For many years the recommendation has been the belt should be fully troughed before the load is introduced. A variation on this thinking is the idea that it is more critical that the belt be stable (i.e., not undergoing transition) when it is loaded than it fully troughed when it is loaded. Given a conveyor where there is a very short distance between the tail pulley and the loading zone, it may be better to partially trough the belt in the area between the tail pulley and the load zone, and then complete the troughing after the belt has been loaded. **(Figure 4.12)** For this to provide any benefit, the belt line must be stabilized with improved support structures, such as impact cradles and side support cradles, and the edges must be effectively sealed after its initial troughing. The final troughing of the belt can be applied after the load is on the belt. For conveyors with inadequate space in the traditional transition area between the tail pulley and the load chute, this method provides the benefit of a higher troughing angle without creating the instability of loading while the belt is undergoing transition.

Gradual Transitions

Another method to deal with the problem of too short a transition distance is the use of a very gradual transition. Rather than risk damage by troughing the belt too quickly, the belt is troughed over an extended distance, making the change so gradual as to be almost unnoticeable. In one case, the belt was changed from flat to 35° trough over a 42-foot (12.8 meter) transfer point. This conveyor was perhaps a special circumstance, with a long transfer point incorporating multiple load zones, a thick belt, and minimal distance between the tail pulley and the first loading zone. The key for this technique is to maintain a straight line for the trough and belt support, without allowing any humps in the belt. Rather than use specially designed components, this gradual troughing change was accomplished by installing conventional components in a "racked" or slightly out-of-alignment fashion. Belt support cradles incorporating sufficient adjustment to accommodate a deliberate "out-of-alignment" installation were used in combination with troughing idlers with adjustable angles.

Increasing Capacity by Dropping the Trough Angle

While in most cases higher troughing angles are used to increase the capacity of the belt, in some circumstances they can have the opposite effect.

To present a "real world" example, the design capacity of a 48-inch (1219 mm) cement plant load-out belt was 1,100 tons (1000 metric tons) per hour. But its actual performance was only 500 tons (450 metric tons) per hour. The belt was troughed at 45°, but this angle, combined with the resulting effective belt width and limited spacing for edge sealing, effectively reduced the carrying capacity of the belt.

Part of the problem was that the finished cement became fluidized as it was transported and, therefore, was not as dense as originally supposed. The existing dimensions were adequate to carry the material in its nominal state, not in the fluidized state. It was a familiar story: you cannot put 10 gallons of material in a 5-gallon pail. A larger chute was needed to allow the specified loading rate.

Lowering the belt's trough angle from 45° to 35° allowed the chute walls to be widened. Moving the chute walls out from a width of 26.5 inches (673 mm) to 38 inches (965 mm) allowed more material to pass through the chute. The lower angle increased the cross-sectional area for cargo, while at the same time providing adequate edge distance for sealing. The result was an overall increase in conveyor capacity to 1300 tons (1180 metric tons) per hour. **(Figure 4.13)**

The effective area of the belt width must be calculated in relation to the troughed angle of the belt and realistic bulk densities of the material in the aerated state must be used.

While this is only one example, it does point out some difficulties in the design of conveyors and transfer points. Many factors must be carefully considered when designing a transfer point. With bulk solids and belt conveyors there are few "givens" and even fewer "constants."

Sealing at the Entry Area

Sealing of the belt entry in the load zone is often a problem. **(Figure 4.14)** The turbulence of material as it is loaded can cause some particles to bounce or roll backward toward the tail of the conveyor. Material will bounce back out of the load zone and roll down the conveyor, accumulating on the pulley, the bearing supports, or on the floor near the tail pulley.

In an attempt to solve this problem, a sealing system of some sort is applied at the back of the loading chute. Typically, this seal is a curtain or wall, fabricated from a sheet of plastic or rubber. But this seal can create as many problems as it solves.

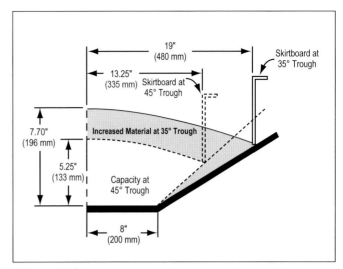

Figure 4.13

By lowering the trough angle, the chute walls could be moved out and capacity increased.

Figure 4.14

The accumulation of material that has rolled back off the tail of this conveyor will soon endanger the belt.

Figure 4.15

If the tail seal is too loose, material will escape out the back of the conveyor.

If the seal at the belt entry into the chutework is too loosely applied, material will escape out the back of the loading zone down the transition area and out onto the floor. **(Figure 4.15)** But if a sealing system is placed tightly enough against the belt to prevent leakage out the back of the loading zone, the seal may instead act as a belt cleaner. In this instance, it will scrape carryback that has adhered to the surface of the belt during the belt's return from the head pulley off the belt. The material removed by this "belt cleaner effect" will then accumulate at the "back door" of the loading zone or roll down the belt to pile up at the tail end of the conveyor. **(Figure 4.16)**

Another problem at the tail end of the chute is difficulty in sealing the corners of the chute, where high material pressures and significant air movement carry dust out of the transfer point.

The difficulty in sealing the entry area is compounded by any dynamic vibrations in the belt created by fluctuations in belt tension resulting from intermittent "peaks and valleys" in material loading or from the use of a wing pulley. Wing-type tail pulleys should be avoided for this reason.

Multiple Barrier Sealing Box

An effective approach is to seal the area behind the load zone with a multiple barrier sealing box. **(Figure 4.17)** This box is attached to the back wall of the loading chute in the transition area between the tail pulley and the first fully troughed idler. **(Figure 4.18)**

A sealing strip is installed on the inside of the back wall–the wall closest to the tail pulley–of the sealing box. **(Figure 4.19)** Deflected by and in the direction of belt

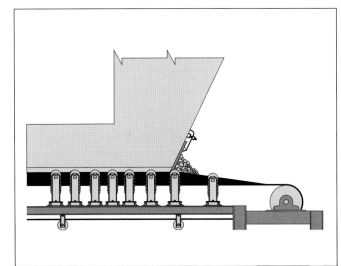

Figure 4.16

(Top) Too much pressure on the sealing strip will remove residual material from the belt.

(Bottom) Belt cleaning effect caused by a high-pressure seal at the back wall of the load zone.

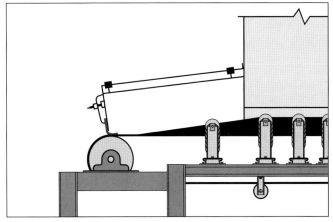

Figure 4.17

A tail sealing box is installed on the back of the chute to prevent material rollback.

Figure 4.18

The tail sealing box is installed between the tail pulley and the back wall of the chute.

motion, this strip forms a one-way seal that prevents material escape. Material cannot roll out the back of the conveyor. As this strip lies on the belt with only gentle pressure, it avoids the belt cleaner effect. Material adhering to the belt can pass under the seal without being "cleaned off the belt." This strip effectively contains material rollback while allowing adhered material to pass through.

On its sides, the box should be fitted with a low-maintenance, multiple-layer skirtboard seal to prevent material spillage over the edges of the belt. **(Figure 4.20)** The tail sealing box should incorporate the start of the conveyor's load zone sealing system, so it runs continuously from the tail sealing box to the end of the skirtboard. **(Figure 4.21)** This continuous seal eliminates the problem of sealing the high-pressure corners of the impact zone.

The top of the tail sealing box shall include an access door to allow the return of any fugitive material to the conveyor. **(Figure 4.22)**

The tail sealing box is most effective when installed where the belt has completed its transition and is fully troughed. This requires less special fitting of the edge seals and results in more effective sealing, but will require some advance planning in the design of the conveyor and transfer point.

Figure 4.20
The sealing box should incorporate the start of the load zone's edge sealing system.

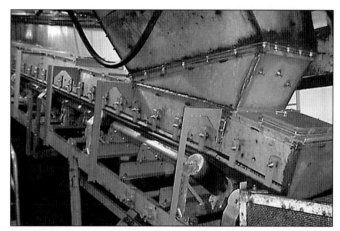

Figure 4.21
The belt edge sealing system should be a continuous strip from the back end of the tail sealing box through the exit end of the skirtboard.

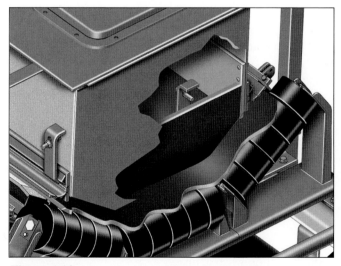

Figure 4.19
A sealing strip is installed on the inside of the box wall closest to the tail pulley.

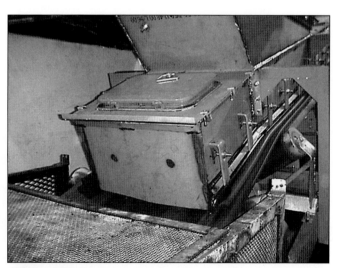

Figure 4.22
The tail box should include a door to allow fugitive material to be placed back on the belt.

chapter 5

Loading Chutes

The conveyor receives its cargo at its loading point(s). This cargo can come from other conveyors or from storage hoppers, process vessels, or transportation systems. While the source of the material may be different, it is all fed to the receiving conveyor through a feeding device called the chute.

The design of a chute is more complicated than just a hollow tube through which material is channeled. The material has its own forces and characteristics, perhaps constant, perhaps constantly changing. Each discharging or feeding structure applies a variety of forces, and each chute adds its own variables of force and friction to the equation.

Consequently, it is important that chutes be properly designed and maintained to minimize the effects of these variables and allow the ultimate success of the loading zone and the entire conveyor.

Problems from Improper Chutes

There are a number of problems that can occur on a conveyor by improperly fitted loading chutes. These include:

1. *Top Cover Damage*
   Turbulent flow of material at the loading point where it is moving too slowly in the direction of the receiving belt can abrade the belt conveyor.
2. *Wear of Skirts*
   Excessive spreading of the material as it lands on the belt results in a large area of contact between the material and skirtboard and, hence, in premature wear of the skirtboard and seal.
3. *Spillage/Dust*
   Poor chute design and construction allow the escape of fugitive materials, which turns into premature equipment failures and high maintenance expenses.
4. *Bottom Cover Damage*
   When spillage occurs, material accumulates on the return side of the belt. This leads to wear of the bottom side of the belt.
5. *Belt Tracking Problems*
   Uneven loading of the conveyor leads to belt wander and, in turn, to damage of the belt edges and even the steel structure.
6. *Impact Idler Damage*
   Forces of the impact from large lumps or long drops onto the belt can damage the belt support structure.
7. *Belt Start-Up Problems*
   A large quantity of material left on the belt when it stops can cause problems when operations resume.

> "The design of a chute is more complicated than just a hollow tube through which material is channeled."

8. *Plugged Chutes*

Chutes can choke and close from material accumulations inside the chute or too small a cross section. Plugs will stop production and can cause severe damage to the chute, belt, pulleys, and accessories. Often the only way to clear a chute is to manually dig it out.

These problems turn into significant costs for materials, for maintenance, and in lost production. But they can be minimized or avoided altogether with proper chute design.

The Jobs of the Chute

It is not possible to detail in a brief passage all the requirements for the engineering of loading chutes. Rather, the following provides an overview of considerations that engineers, maintenance crews, and operations personnel will need to evaluate an existing chute or chute design; and ascertain how its performance will affect the transfer point, the conveyor, and the overall operation.

Transfer point chutes have six missions:
1. To feed the receiving conveyor in its direction of travel.
2. To centralize the material load.
3. To minimize impact on the receiving belt.
4. To supply the stream of material at a speed equal to the speed of the receiving conveyor.
5. To return belt scrapings to the main material flow.
6. To minimize the generation and release of dust.

The above should not be considered absolute requirements but rather goals. These various responsibilities may conflict with the realities of industrial settings and real-life conditions. However, these six tasks together can be considered the ideals for a chute, and the closer any chute comes to balancing and achieving all six, the more satisfactory will be its performance.

Mission #1 "In The Direction of Its Travel"

Ideally, all transfer points would be in-line. That is, two belts would be running in the same direction, hence, it would be relatively easy to place the material on the receiving belt in the direction it is to be carried. In-line transfers are incorporated into systems to reduce the length of a conveyor when insufficient power or tension is available with a single belt, or to accommodate a mechanism for material separation or another process step.

More typically, however, a change in direction is required, and this is the reason (or at least one of the reasons) the transfer point is installed.

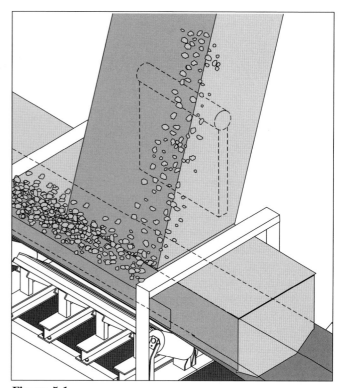

Figure 5.1
Deflectors are placed within chutes to control load placement, impact, and wear.

There are a number of drawbacks to angled transfer points. With these transfer points it is more difficult to maintain the angle and trajectory of the material, the speed of the material, and the service life of components in the transfer point.

Despite these drawbacks, there are reasons that angular transfers might be included in an operation's flow plan. Angle transfers may be needed to accommodate directional changes in material flow due to site restrictions. In other cases, the angular transfer may be included to divert or split the main stream for stockpiling or material separation.

There are a number of strategies and components that can be employed to direct the material flow in non-in-line transfer points.

Deflectors

Many chutes–especially those used in non-in-line transfer points–require positioning of one or more deflectors or impact plates to retard the material's forward momentum and redirect it into the direction of the receiving belt. These devices steer the material flow; and in so doing, allow the designer to limit the physical dimensions of the chute to a size that is manageable, given the available space and budget. The high speeds and long discharge trajectories of modern conveyors would require larger and more expensive chutes to contain the material flow. The use of impact plates and deflectors help control the material passage within reasonable limits.

Sometimes deflectors are used inside a chute to absorb impact and minimize wear. **(Figure 5.1)** Other times, they turn the moving material, forming a sacrificial wear part to redirect the flow, and preserving the chute wall and the belt from premature wear.

Curved plates–sometimes called spoons–are at times used at the bottom of the loading chute to steer material and minimize its spread. Curved plates coax the stream to exit at the required angle and tend to reduce material spread and degradation. **(Figure 5.2)** Without this curve in the loading plate, the material may be deposited in a direction that does not coincide with the receiving belt's direction of travel, leading to turbulence, pooling, and possible belt mistracking. Care must be exercised with curved loading plates to make sure they are self-clearing, as to prevent material accumulation leading to blockages of the chute. Caution should be used to maintain enough clearance between the deflector and the head pulley of the discharging conveyor. If the material is very cohesive, it can adhere to the plate (and to itself) well enough to bridge the gap and plug the chute. This problem is worse with slower moving streams, where the "scouring effect" of following material is reduced, so that material can build up at higher rates.

Another problem is that material adhering to the plate will change the "throw" of the deflector. Deflected material will then go in different, unplanned directions to strike unprotected walls or land on the belt with a different "spin," so it is no longer moving in the correct direction.

The installed plates should be readily field-adjustable, so they can be positioned precisely (both at installation and after wear) to achieve the desired effect. These deflectors should be accessible to allow fine tuning and efficient replacement.

Inspection and access points are critical to maintaining the proper direction for deflected material.

Mission #2 "Centralizing the Load"

One of the most important missions of the chute and one critical to control of spillage is the placement of the load in the center of the belt.

Off-center loading–placing the material predominantly on one side of the belt–is a problem at many transfer points. It is most commonly seen on transfer points that must change the direction of material movement. It is also seen with in-line transfer points. This is true particularly when material accumulations within the chute, changes in material characteristics (such as moisture content, particle size, or changes in material speed during start up or shutdown), alters the trajectory, and piles material deeper on one side of the conveyor belt.

Problems from Off-Center Loading

Off-center loading causes numerous operational and maintenance problems that may manifest themselves in unusual ways.

When material is not spread uniformly on the belt, but rather piles up against one of the skirtboards, the belt can be displaced transversely. **(Figure 5.3)** When the belt is loaded off-center, the center of gravity of the load shifts as the material tries to find the center of the troughing idlers and cradles. This forces the belt off track, pushing it toward its lightly loaded side. This displacement causes belt training problems and may result in spillage over the edge of the belt outside the transfer point.

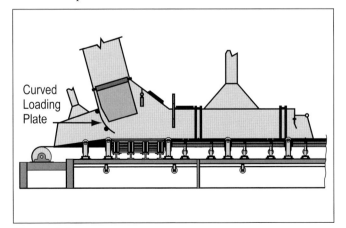

Figure 5.2

Curved plates at chute bottoms reduce material speed, impact, and degradation.

Figure 5.3

Off-center loading will push the belt to one side of the loading zone.

If the material is loaded in a direction not in line with the receiving belt, it is likely that wear patterns will soon become visible on the inside of the chute. These patterns will correspond to the path the material takes as it ricochets off the inside of the chutework, trying to reach equilibrium with the direction and speed of the belt. It is common to see this wear repeated two or three times within a transfer point, and yet not be visible in the appearance of the load on the belt as it exits the skirted area. This movement of the material, as it bounces back and forth within the transfer point, will accelerate wear on the skirts and liners. Once there has been enough wear to create a cavity where material can lodge, entrapment begins, and the wear rate accelerates. Although these cavities may be only 12 inches (300 mm) long, the forces of the moving material quickly put tremendous pressure onto the skirting seals. In some cases this is why transfer point skirting seals will leak on one side only. In other cases, the belt moves all the way out from under the skirting on one side. The sealing strip then drops down and blocks the belt from returning to the centered position. This continues until the seal is sawed off from the belt edge running against it. Of course this creates a significant spillage problem.

It is important to anticipate and correct these problems. Training idlers and other belt aligning systems *(as discussed in Chapter 15)* provide only limited success in compensating for the effects of off-center loading. It is more effective to install the corrective measures within the loading zone.

Curing Off-Center Loading

The only sure cure for off-center loading is a properly designed and engineered feed chute or loading hopper. The use of properly tuned deflectors, liners, baffles, shapers, screens, grizzly bars, or curved gates within the chute helps direct the flow of material to provide a balanced loading pattern. Alterations in the loading gates or chutework should be plotted to allow the centering of the load regardless of the load's size or consistency.

One option is to install adjustable gates that are trained to direct the material path into the center of the receiving belt. These moveable plates should be adjustable from outside the chute while the system is operating, allowing easy correction of flow should material characteristics or flow rate change. One manufacturer offers a self-centering loading chute. Whether controlled automatically or manually, load centralizers should be designed for easy maintenance or replacement, as they will be subject to considerable wear from material abrasion.

The discharge trajectory from belt feeders can be difficult to center, because they typically operate at a variable speed, which, of course, alters the discharge trajectory. Increasing the drop height creates more room to allow the material to be directed. It is possible to provide an automatic system where devices such as belt alignment switches or limit switches control the placement of a moveable gate. However, directing the material onto the belt in the same direction and speed that the receiving belt is moving is usually the best solution.

Sometimes with an angled (non-in-line) transfer, the load may be centered, but the force of the material trajectory can push the belt off center. In this case, it is better to revise the chutework so the trajectory is in line with the belt travel. This may be difficult at times because of headroom restrictions. The only alternative may be to change the trajectory to a vertical drop, eliminating all lateral or horizontal thrust from the load. Doing so may reduce the life of the belt covers, particularly if speeds are high, and material is coarse or sharp-edged.

Load Centralizers

Specialized systems that form the belt into a deeply troughed U-configuration to force the load back into the center are available. Some of these systems also incorporate a "belt beater" or vibrator to encourage the shift of the material. Load centralizers represent an option that might be selected to cure one problem, but creates undesirable side effects.

Load centralizers can be used effectively if there are multiple load points, found on the main haulage belts in underground mines or on a conveyor belt that goes beneath a storage pile with several draw-off points. It is easier to maintain an effective seal on such a conveyor if the load is centered. It is also easier to hold the skirting and sealing rubber close to the belt edge at a load point if the approaching load already on the belt does not splash against the sealing hardware. A centralizer located 20 to 30 feet (6 to 9 meters) past each loading point will positively center the load, making it easier to add another layer of material without fear of spillage.

If the centralizer needs to vibrate the belt in order to center the load, there is the possibility that this oscillation will be transmitted back into the loading point, making it more difficult to seal this area. Therefore, proper belt support should be installed through the sealing area.

Catenary idlers, with their ability to reshape themselves under the forces of loading, may also provide a load centralizing effect, although their overall flexibility makes them very difficult to seal against effectively.

Deflector Liners

Deflector wear liners inside the chute can sometimes be useful in reducing the problems from off-center

loading. These liners feature a bend or angle that turns material toward the center of the belt, away from the belt edges.

Uneven wear on these (or any) liners can create entrapment points, which lead to abrasion of the belt and material spillage. The worn area tends to trap particles that pinch between the belt and the wear liner or between the wear liner and skirting seal. This trapped material quickly wears a narrow groove into the belt. This groove is almost always "blamed" on the elastomer skirt-sealing strip being too hard or of an abrasion rating greater than the belt. In fact, it is very rare to find this wear directly attributable to the seal material. Rather, this wear originates in the abrasion of the belt by the entrapped material. Periodic inspections to spot the entrapment points as they develop and planned maintenance to correct these gaps is a must to reduce fugitive material and maximize belt life.

As noted elsewhere in this volume, deflector wear liners create other problems such as material entrapment and chute choking. These drawbacks may be so severe as to prevent these liners from being an effective measure for centralizing material on the belt.

Flow Training Gates

The chute can help control load placement through the use of adjustable flow training gates or shapers. These are wedge-shaped deflectors installed on the inside surface of the chute. They serve to direct lumps of material toward the center of the load zone where they are less likely to slip off the edges of the belt or damage the skirtboard seals.

It is common practice during the initial start up of a new conveyor system to install deflectors within the chute to center the load. All too often, however, no permanent record is kept of the position or size of these deflectors. As they wear, problems begin to show. Meanwhile the project has been accepted, the contract has been closed, and the people who designed the deflectors and made them work at start up are nowhere to be found. This creates another period of trial and error for the design and positioning of replacement deflectors, a period with increased spillage and higher risk of component damage. It is important to keep detailed dimensional information on deflectors, so they can be duplicated when replacement is required.

Mission #3 "Minimizing Impact"

Loading zone impact comes from short drops of large-sized lumps or the long drops of quantities of small-sized material. **(Figure 5.4)**

In an ideal world, the height of drop for the material onto a belt should not be more than 1 to 1.5 times belt width. However, practical considerations–ranging from

Figure 5.4

(Top) Impact in the loading zone may originate in long material drops.

(Bottom) The loading of large lumps of material may cause high-impact levels.

the space requirements of crushers to engineering changes in process, equipment, and materials–often mandate drops of far greater distances.

Headroom is sometimes kept to a minimum in an attempt to keep down the overall height (and thus the cost) of a conveyor structure. As a structure goes higher, it requires additional steel for support, and the system requires additional energy to drive the belt up the incline. As a result, designers try to keep the height of transfer points to a minimum, so the head pulley drops the material directly onto the next belt. But the closer the two belts are together, the less room there is for adequate belt cleaning systems. So in minimizing one problem (impact) another problem (carryback) is worsened.

The challenge is to create transfers that provide enough space between conveyors to allow equipment installation and maintenance, yet control drop height and impact levels.

Chute Angle and Material Trajectory

The angle at which the chute descends from the unloading structure onto the receiving belt should be flat enough to keep lumps from bouncing excessively after they land on the belt. A chute with a low angle, combined with proper load direction and speed, causes the lumps to strike the belt at a grazing angle. This allows the material to bounce more gently, carried along in the direction of belt movement, rather than to rebound back into the face of the incoming material stream. This reduces the risk of damage to the belt as well as minimizes material degradation and dust generation. But if the chute angle is too flat, material flow will slow, leading to a clogged chute.

The effort, therefore, is to provide a chute angle that moves material at the appropriate speed and volume, providing efficient loading without unusual wear or abrasion and without increasing impact damage or dust generation. As always, the consideration of material should include an understanding of its "worst case" characteristics.

The calculation of this chute angle should be based on a study of the material to be handled plus an appropriate allowance, depending on whether the material is in a relatively static condition (being loaded from a hopper) or in motion (coming from another belt). Typically in real life, this calculated "ideal" angle will be compromised by other needs of the conveyor and the overall process.

Material on Material

One of the most popular ways to minimize impact is to confine the impact forces as much as possible to the material itself; to use the material itself to absorb the impact and abrasion of subsequent material.

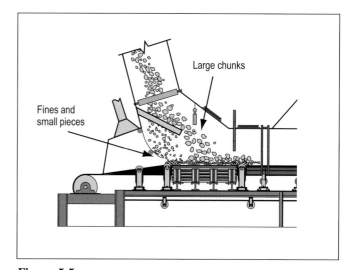

Figure 5.5
Fines sift through grizzly bars to form a bed of material on the belt and cushion the impact of larger lumps.

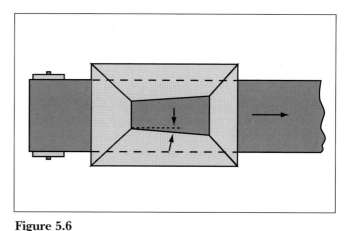

Figure 5.6
A wedge-shaped loading chute allows smaller particles to load the belt before the larger pieces.

Material-on-material techniques include the installation of scalping or grizzly bars in the chute. **(Figure 5.5)** These consist of spaced bars that allow the fines to pass through first to form a protective bed on the belt. The lumps, unable to pass between the bars, slide down the incline and land on the belt on a cushion formed by the previously deposited fines.

Wedge-shaped loading chutes serve a similar purpose by letting a layer of fines pass onto the belt in advance of any larger lumps of material. (**Figure 5.6**)

Rock Boxes

Another method to minimize wear through material-on-material impact is to incorporate a "rock box" inside the chute to create a pattern of "cascade transfer." **(Figure 5.7)** Rock boxes begin with a ledge inside the chute. A pile of material accumulates on this shelf. Subsequent material moving through the chute flows over or deflects off this pocket of captive material. Abrasive force is thereby shifted from the chutework to the accumulated bed of material. Impact force is also dissipated as material bounces off the material on the ledge. As the chute is now converted from one longer drop to two (or more) shorter drops, the level of impact in the loading zone and on the belt is reduced. Induced air movement in the transfer point is also reduced, which provides benefits in the management of airborne dust.

When handling fine or abrasive materials, a number of small steel angles can be welded onto the chute bottom to form mini-rock boxes. These smaller rock boxes allow oncoming material to slide over the material captured on the ledges, reducing wear and abrasion on the chute wall and bottom.

Similar to rock boxes are rock ladders, which are composed of a series of baffles or mini-rock boxes. (**Figure 5.8**) The shelves are typically arranged on alternating sides of the chute, so the discharging material never has a free drop of more than five to six feet (1.5 to 1.8 meters). The effect is to reduce impact on the belt, as well as control material fracture, dust creation, and air movement.

Difficulties with Rock Boxes

Rock boxes, when carefully incorporated into the design of a transfer chute and trimmed to suit on-site conditions, can be successful in reducing impact wear. They are most successful if requirements and physical conditions and flow rates do not change over time. However, these factors do not remain constant in many operations.

It is important that the material continue to scour across the rock box's buildup. Care must be taken to accurately judge the cohesive characteristics of the material (under wet conditions, for example) in order to avoid accumulations that can choke the chute.

The face of the compacted material body in the box is not flat, although this is typically assumed by the designer. The impact surface is often curved; shape and angle of incline will vary with the cohesiveness and internal friction characteristics of the material. The slope angle can range from 30° to 90° (which could block the chute). In addition, tramp iron can become trapped in the box, adding the risk of ripping the belt or altering the flow pattern. The prudent designer must increase the distance of drop to the ledge to ensure accumulated material does not build up high enough to foul the head pulley, belt cleaners, or dribble chute of the discharging conveyor.

After the material hits on the inclined face of the rock box surface, the material outflow has a horizontal component that is unpredictable and may not be in the direction required. This is especially the case in non-in-line transfer points. Even if the outflow pattern of the box is adjusted during the start up, random changes in material characteristics, the evolving geometry of the wear components, and even changes in the load of material on the rock box ledge can create problems in predicting flow direction.

Rock boxes should not be used in transfer points handling fragile bulk solids that might suffer degradation or materials with large lumps that can block or choke the flow. Rock box designs should not be used if a conveyor will carry more than one material.

Impact Plates and Grids

Another method to divert flow and absorb impact is the use of plates or grids in the material path. An

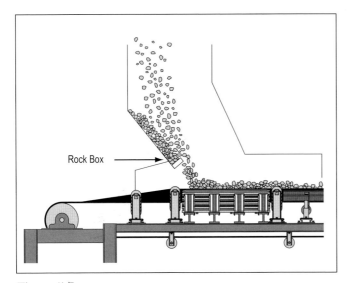

Figure 5.7

Rock boxes are installed inside chutes to create a material-on-material zone and minimize impact in the load zone.

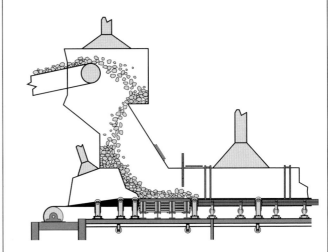

Figure 5.8

To break up longer material drops, multiple rock boxes can be used to form a rock ladder.

impact plate is a consumable barrier that is placed inside the chute to absorb force from the moving material stream. This is especially important where angled transfers and moderate-to-high belt speeds are present, and circumstances (such as available space and budgets) prevent the engineering of ample chutes.

Rock or impact grids catch material to develop a material-on-material impact that preserves the chute walls. Subsequent material bounces off the captured material without actually hitting the grid or the chute wall. **(Figure 5.9)** The grids are mounted on the walls of the chute to absorb the horizontal velocity of the discharging material trajectory.

To accommodate the occasional oversize rock or tramp material, the gap between the head pulley and the impact plate or grid should be carefully engineered. This will also minimize problems due to cohesive or high-moisture materials that can lead to choking of the chute.

Flat impact plates can be used where belt speed is slow enough that the trajectory of the material discharge hits the plate well below the pulley. When the belt is moving faster so the discharge hits the plate above the pulley, a significant amount of material will be supported on the following stream. The supported material will slide to the outside, where it will build up on the walls and eventually plug the chute. Adding angled sidewalls to the impact plate will solve this problem by allowing gravity to force the material to fall.

The selection of appropriate materials and careful attention to design and positioning can provide significant improvements in the life of these wear components.

Mission #4 "Equal to the Speed of the Receiving Conveyor"

When a slow-moving load hits a fast-moving belt, there is a slipping action that causes substantial abrasion of the receiving belt's cover. By loading material at or near belt speed, this slip is reduced and belt life is improved.

The use of short transition or "speed up" belts running in the same direction and speed as the main conveyors is one method of absorbing the punishment of this impact and wear. **(Figure 5.10)** These shorter belts guide the material onto the longer conveyor and accelerate the material in the desired direction. The shorter transition belt is sacrificed by making it take the abuse of impact and of changes in speed and direction. This spares the longer, more expensive belt from these damaging forces.

The Problem with Pooling

Another phenomenon that occurs at transfer points where material falls vertically (or nearly vertically) onto a high-speed belt, is pooling. Material that is not yet moving at belt speed piles up on the belt over a length of the conveyor, creating a "pool" of material in the loading zone.

When material falls nearly vertically, the receiving belt is running more-or-less perpendicular to the motion of the falling material. Under these circumstances, the loading material is brought up to belt speed through its contact with the receiving belt. Particles or lumps of material drop onto the belt and then bounce and tumble, dissipating the energy supplied by the previous conveyor and the energy from its impact with other lumps, until it can be caught by the motion of the receiving belt.

The greater the differential between the speed of the receiving belt and the velocity of the material stream, the longer and deeper the pool of material. As this body

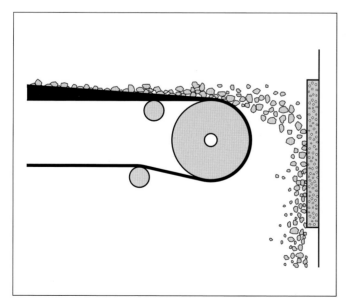

Figure 5.9
An impact grid acquires a layer of material, preventing wear in conveyor structures.

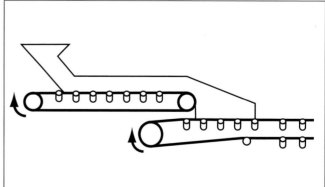

Figure 5.10
Short "speedup" conveyors can be used to minimize wear on a longer, more expensive belt.

of material grows, it becomes more difficult to maintain a sealed, spillage-free transfer point.

A speed up conveyor can be used to remedy this condition. Or it might be practical to use a curved gate or ramp to manipulate the direction and velocity of the loading stream until it more closely resembles the speed and direction of the receiving belt. These curved chute bottoms steer the material flow to place it on the receiving belt centrally, moving in the proper direction at the proper speed with low impact levels. This more careful and gentle positioning of the load on the receiving belt leaves less energy available to generate dust.

Mission #5 "Returning Belt Cleanings"

Belt cleaners are installed at the head pulley to remove residual material that has adhered to the belt past the discharge point. The material removed by cleaners should be returned to the main material body and not allowed to build up on chute walls or other components. Consequently, the chute should be large enough and with steep enough walls, so removed material can be directed back into the main material flow. This mission may require use of oversize chutes, low-friction chute liners, or auxiliary devices such as vibrating dribble chutes and scavenger conveyors.

Belt cleaners and dribble chutes are discussed in greater detail in Chapter 13: Belt Cleaning Systems.

Mission #6 "Minimizing Dust"

In addition to controlling carryback and spillage, the well-engineered and constructed chute can make a significant contribution to management of airborne dust.

The chute and the skirtboard sections should be large enough to provide a plenum that minimizes air currents and reduces the positive pressures that can carry material out of the enclosure. While these steel structures are off the surface of the belt far enough to provide for the belt's safety, they should fit close enough to the belt to permit an effective seal.

The enclosure should be spacious enough to permit internal circulation of the dust-laden air. A large loading zone chamber allows dissipation of the energy that carries off the dust. It is important that the material fines be allowed a chance to settle, either on their own or with the addition of a water spray or fog system, before dust take-offs are installed. Otherwise, energy will be wasted suppressing or collecting dust, which would have shortly settled on its own. This would cause the dust suppression and collection systems to be larger (and more expensive) than would otherwise be required.

One or more access points into the enclosure should be provided to allow for service, but these ports should

Figure 5.11
Holes in chutework can allow significant amounts of material to escape.

Figure 5.12
Patches or "scab plates" can be used to cover holes in chute walls.

be fitted with doors with tightly fitting covers to prevent material escape.

Even the smallest openings should be considered a source for dust. Load chutes with holes from rust or abrasion can allow significant amounts of material to escape. **(Figure 5.11)** Access doors that are not completely closed can allow significant amounts of dust out. Even the holes created by missing bolts can vent a visible stream of airborne dust. For total dust control, these openings must be closed with caulk, a "scab plate" **(Figure 5.12)**, or a properly designed replacement chute.

Rubberized fabric can be stretched over the mounting supports enclosing plant equipment to provide another method to control dust arising from leaking chutes and non-covered skirtboards. Systems providing durable fabric with easy-to-use "grip strips" that clamp over a beaded edge installed on the equipment are available. **(Figure 5.13)**

Physical and Computer Modeling

Now it is possible to model, both physically and electronically, the design for a chute. These procedures allow the development of new designs and the alteration of existing systems.

Physical modeling involves construction of a scaled model of the transfer chute and simulating the flow of material through that chute. The majority of these models have been conducted with a 1:10 scale. A smaller scaling factor (i.e., larger model) will cause difficulties in attaining the required tonnages on the conveyors currently available. Choosing a larger scaling factor (i.e., a smaller model) is typically not possible due to limitations in the ability to provide accurately sized particles and lumps.

This technique is beneficial as it can provide a reliable and cost-efficient means for addressing problems with transfer chutes. However, the problem with the physical modeling is that it requires the ability to match lump size and cohesiveness to get a true reading on the performance of the material and of the chute.

Software has now been developed that models how materials flow in a conveyor system. It can even provide "what if" analysis, allowing for unusual events or circumstances. Mathematical design processes integrated into the modeling program can generate detailed drawings. The basis of the process is parametric design that can be done in six to eight hours following the detailing of the key 20 to 25 parameters by plant personnel.

The first fundamental requirement is modelling the material trajectory from the pulley face of a discharging conveyor (or other structure). This is calculated from input including:

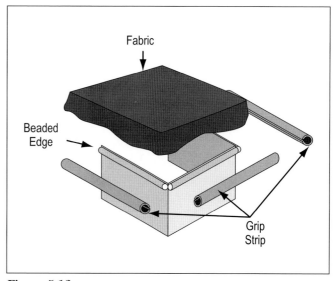

Figure 5.13
Fabric held in place with removable strips can prevent fugitive material.

- Belt speed
- Belt width
- Material size
- Material composition and physical properties
- Angle of inclination of the discharge conveyor
- Pulley diameter
- Transition arrangement and length

After the trajectory is plotted, the software focuses on handling the energy in the material by designing a receiving spoon and a corresponding covering hood.

The process develops a computer-generated and rendered color model of the transfer assembly, including structural steel. The model can be made to simulate dynamic materials so that specific problems such as large irregular lumps can be assessed. This allows for the checking of designs before construction.

This computer model can be turned into fabrication drawings, 3-D installation drawings or even the bill of materials needed for construction. As this methodology is computer-based, it becomes relatively easy to incorporate changes and evaluate their impact on the chute and system as a whole.

Again, this technique offers some assistance in developing solutions to chute problems. It seems readily applicable to new conveyor construction. However, its application to the retrofit of existing chutes or new conveyors installed in existing spaces is more problematical.

"Hood and Spoon" Chutes

This computer modeling of flow is an important ingredient in the development of "hood and spoon" chutes. Each "hood and spoon" chute must be engineered to match the application, with material characteristics, flow requirements, and conveyor specifications, all ingredients in determining the proper design.

Properly matched to the system, "hood and spoon" chutes will control the movement of material to prevent dust and fugitive material and reduce belt damage. They feature two components that confine the material stream to provide an efficient, low-dust transfer. **(Figure 5.14)**

Installed in the material trajectory of the discharging conveyor, the "hood" serves to receive and deflect the material downward. Rather than letting the material stream expand and pull in air, it keeps the stream (and the individual material particles) close together. There is less air captured that will be released when the material lands on the belt. With this reduction in the air carried with the material stream, there is less air (and airborne dust) expelled when the material lands on the receiving conveyor.

The "spoon" forms a curved loading chute, like the slide on a schoolyard playground. This forms a smooth line of descent so that material flows evenly onto the receiving belt. The spoon is designed to lay the material on the belt at the proper direction and speed. This will prevent turbulence, reducing material "splash" out the back of the loading zone and minimizing impact (and the damage and spillage it can cause).

When combined with effective seals at the belt entry to discharge and at the exit of the receiving conveyor, this system has been shown to provide effective dust control.

Chute Dimensions

A loading chute that is too wide for the belt and the rest of its equipment in the transfer area may not allow sufficient room for lateral belt movement or for the installation of effective sealing systems. The CEMA book *Belt Conveyors for Bulk Materials* recommends (and other publications reiterate) that the loading chute width should be no greater than two-thirds the width of the receiving belt. From a practical standpoint, wider belts can go beyond the two-thirds rule so long as there is enough belt edge outside the chute wall to accommodate sealing systems and to allow some lateral movement of the belt. A six-inch (150-mm) belt edge is more than adequate. On very narrow belts, the two-thirds rule does not allow sufficient belt edge for effective sealing. *Skirtboard Width is discussed in greater detail in Chapter 7.*

CEMA also specifies that when lumps and fines are mixed, the inside width of the chute should be two times the maximum lump size.

The volume (or cross section) of the chute should be at least four times the load stream from the feed conveyor (or hopper). This provides sufficient margin

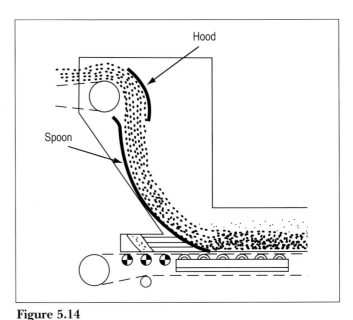

Figure 5.14
A "hood" confines the material stream to minimize induced air; a "spoon" loads the receiving belt in the proper direction and speed with low impact.

to minimize the chance that the chute will become blocked by an arch of accumulated material or become full during belt stoppages.

In addition to determining chute size, these specifications may also provide insight into the selection of the size of the receiving belt. A common problem that cannot be rectified by chute redesign is the operation of a belt over capacity. There are a number of benefits to installing a belt wider and running it slower than strictly necessary to carry the design material load. One major benefit is a reduction in transfer point problems such as spillage and mistracking. There will be less material out to the edges of the wider belt. A wider, slower belt also has potential for later upgrades in capacity (through increasing its speed) than a conveyor that is engineered at the upper limits of its capacity.

Chute Construction

Chutes are generally fabricated from steel plates, typically mild or stainless steel, depending on the material that is to move through it. Reinforcement beams are of similar steel as required.

The selection of plate thickness is dependent on the nature and volume of material moving through the chute, the structural strength requirements, and the margin for wear if the chute is not to be fitted with a replaceable liner system.

Chutes are typically engineered to be fabricated in sections of a size that is convenient for transport, and then erected on site. One result of this procedure–due to field conditions and on-the-spot decisions–is the engineering drawings of the chute may not be accurate.

Another common practice is the hanging of steel plate on an installed framework with the drawings done "after the fact." Neither practice lends itself to a precise record of equipment design or precise control of material flow.

Whenever possible, conveyor chutework should be installed under the direct supervision of, or better yet by, the company that did the engineering work. When multiple workforces are used, each crew makes decisions based on what is expedient for themselves, rather than on what will result in the best conveyor system. It is not uncommon to see chutes designed to be one-half inch above the belt actually installed at three to four inches off the belt, to make stringing the belt easier. But in consequence, this critical dimension that would have helped control fugitive material has been abandoned.

The chute can be bolted or welded together. Bolted chutes trade off the advantage of easier assembly or disassembly with the disadvantages that bolted joints are more likely to leak material (due to improper sealing/gasketing), and the protrusions of the nuts and bolts provide a ready target for material impact, abrasion, and buildup.

Care must be used in the construction of chutes to avoid imperfections in the surface that can disrupt the material flow and negate the careful engineering that went into the design. Even small variations of ±1/8 inch (± 3 mm) can be too much when matching wear liner sections or truing the chutework to the belt. Precision and care during chute installation will be returned many times over in improved efficiency and in ease of maintenance.

The care of the steelwork presents a continuous problem as wear, age, and "band-aid" repair work gradually alter the original configuration. These patches can create ledges and pockets to catch material and disrupt flow. Again, care in maintaining the original contours will assist in retaining "like new" performance.

Lining the Chute

Wear is an area that can only be brought under control and not altogether eliminated. The continual bombardment of material against the sides of the chute and on the belt is the main source of wear in a transfer chute. The cost of replacing worn parts will always have to be borne, but can be minimized by careful design.

One way to minimize wear is the use of sacrificial liners inside the chute. Liners can also be installed for the specific purpose of reducing wall friction and adhesion. In addition to the grids and impact plates discussed above, various types of linings are available for use inside the chute. In selecting a material for use as a liner, a compromise has to be struck between materials that will resist abrasion and give satisfactory life, and those that have acceptably low wall friction to minimize blockages.

Keep in mind that the rougher the surface–originally, or through corrosion, wear, or abuse (like hammer marks)–the steeper the angle required to keep material moving in a chute.

Materials available include ultra-high-molecular-weight polyethylene (commonly called UHMW), ceramic tiles, and urethanes, as well as mild steel, stainless steel, and abrasion resistant plate.

As non-steel linings are installed, it is important to consider the different expansion rates between linings and the steel surfaces to which they are attached. If there are large temperature changes, gaps must be left between adjacent pieces. Typically, an expansion joint of 1/8 inch (3 mm) is allowed every two to three feet (600 to 900 mm).

Ceramic Linings

Ceramic tiles or bricks can be glued or welded (or both) into place. Ceramic tiles are usually a good choice in high-temperature conditions or in applications that are "hard to get at," as they offer durability and low maintenance requirements. Putting in an "expensive" ceramic liner that could last many years can save money over the long term.

UHMW

UHMW (Ultra-High-Molecular-Weight Polyethylene) is available as new or reground material with a variety of optional additives to improve performance. These additives would include glass, sand, or silica to stiffen and improve thermal stability, silicone to improve lubricity, and carbon black to prevent degradation due to sun or ultraviolet light. UHMW is typically installed with bolts or with adhesive.

Urethane

Two varieties are available. Polyether urethane, while more expensive, offers better abrasion resistance and "wet" condition performance. Polyester urethane has better cut/tear resistance and better impact resistance. Urethane is not generally recommended for high-temperature applications or severe impact areas.

Use of Flow Aid Devices on Chutes

Despite the best intentions of designers, there are occasions where there will be material buildup in transfer chutes. Materials with high-moisture content can adhere to walls or even freeze in the winter operations. Continuous operation can serve to compress the material encrustation harder and more firmly onto the wall. In some cases, the chute will be completely blocked.

The old solution for breaking loose blockages and removing accumulations from vessels was to hammer on the outside of the chute. But this proves a self-defeating program, as the bumps and ridges left in the wall by the hammer blows leads to additional material accumulations.

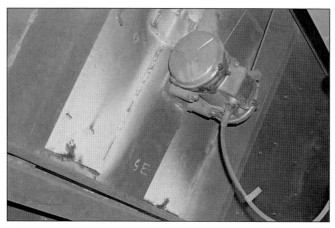

Figure 5.15
Installation of a vibrator can maintain flow through transfer chutes.

Figure 5.16
Air cannons can be installed to prevent buildups and blockages in chutes.

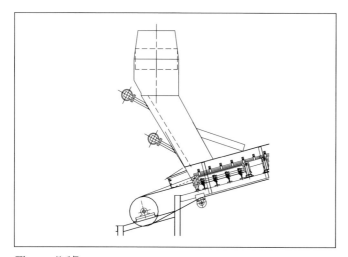

Figure 5.17
The discharge of air cannons down the valley angles of a chute will remove material accumulations.

A better solution is the application of a vibrator on the exterior of the chute. These devices supply energy that will reduce the friction of the walls and the material to keep the material sliding to the chute discharge.

Most commonly, a single vibrator is applied on the wall that features the shallowest angle or is the most buildup-prone. **(Figure 5.15)** Both air and electric vibrators have been used, depending on the available sources of power.

Combating Blockages with Air Cannons

Another successful solution to chute buildups is the use of air cannons on transfer point chutes. **(Figure 5.16)** Two installation techniques have been developed for this application.

One involves using flat nozzles to release the "blast" from the cannon directly into the material. Specially-designed flat nozzles are installed into the chute wall where buildups are found.

The number of air cannons required depends on the size and shape of the chute. Air cannons are installed at several heights around the vessel, and the discharge firing cycle works from top to bottom of the installation. **(Figure 5.17)**

The nozzles are embedded in the wall, so they discharge under the encrustation of material. Care must be taken when installing the nozzles so as to not create additional edges and corners that allow material buildup. The movement of the bulk material will wear the nozzles and, with large lump sizes, deform or even destroy the nozzles.

The second installation technique has the air cannon discharge into the back of a flexible lining blanket or lining on the chute wall. **(Figure 5.18)** This gives the blanket a periodic "kick" to break loose the material, like shaking sand out of a blanket at the beach.

Normally, the blanket is installed on only one side of the vessel, the flatter or less free-flowing side. This blanket is secured only at the top.

The discharge outlet of the air cannon is covered completely by the curtain. The blowpipe is positioned in a horizontal direction or inclined down from the cannon to the outlet to prevent material from entering the blowpipe and valve.

Typically, one air cannon can keep 15 to 20 square feet (1.5 to 2 square meters) of chute wall free of material. Air cannons with a volume of 1.75 cubic feet (50 liters) of air and a 4-inch (100-mm) discharge show the best results.

The advantages of the blanket installation are reduced abrasive wear on cannon outlet and less dust up the discharge. The one-half-inch (12 mm) thick blanket will see some abrasion from the material.

Regardless of technique used, the firing cycle for the air cannon installation must be adjusted for the specific conditions of material, chute and climate. After satisfactory results are obtained, the cannons can be put on an automatic timer, so their discharge cycle continues removing material without attention or intervention from plant personnel.

Discharge of the air cannon into the vessel causes a dramatic increase of positive pressure within the chute, and can increase the release of dust from the chute or loading zone. However, in many cases air cannons are used on sticky materials that are not prone to high levels of dust generation.

It is critical that the steel chute and support structure are sound, as the discharge of one or more air cannons into this vessel can create potentially damaging vibrations.

Chute Access

It is important that any enclosed chute be provided with adequate access. Not only must openings be installed–both inspection windows and worker entry ports–but there must be paths cleared for workers to reach these openings. *Access requirements are covered at greater length in Chapter 18: Access.*

Inspection openings should be positioned away from the flow of material and placed to allow observation of the material movement and checks for wear at critical areas. Covers should provide a dust-tight seal and resist corrosion. Screens, bars, or other obstructions should be in place to prevent material from being cast out of the chute and personnel from reaching into the material trajectory.

It is essential that liner replacement be made as "painless" as possible, as plant personnel tend to avoid this necessary service. For example, the maintenance crew may decide not to "change out" a liner because "it is only worn an inch off the bottom." But this inch is the critical factor in the life of many transfer point components, as the open space allows development of an entrapment point that can fill with material and gouge the belt or abuse the rubber seal.

Often forgotten in the design of transfer chutes is the provision for some method of access to replace liners inside the chute. It is embarrassing to design a small chute with replaceable liners only to discover later that once the chute is in place, there is no way (or at least no efficient way) into the chute to remove the worn liners or to put new sections into position. Chute liners are heavy, no matter what type is used, and they can be hard to manipulate when inside the chute. Normally, some kind of hinged access door at the back of the chute is required to allow both inspection and replacement of the liner.

The consideration of service is particularly important on chutes too small to allow personnel to work inside. Fabricating these chutes in sections for easy disassembly is one approach. Another option would be the flange bolting of the non-wearing side of the chute to allow the opening or removal of the entire side panel.

Safety considerations require that access be limited so that personnel cannot enter the chute until appropriate safety procedures (i.e., locking out of the conveyor drive) can take place. No one should enter these confined spaces without proper training and safety procedures.

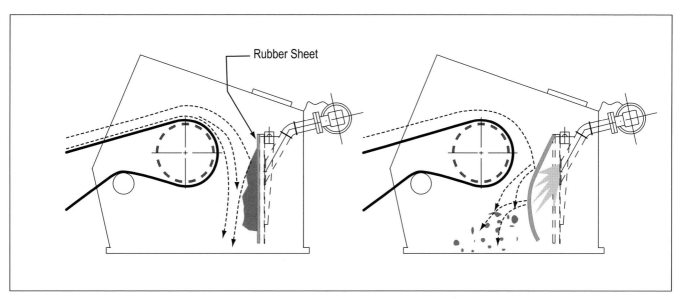

Figure 5.18
The discharge of an air cannon into a rubber blanket will keep material from accumulating on the vessel wall.

Chapter 6

Belt Support

"...the design engineer should do whatever possible to keep the belt's travel line consistently steady and straight."

The construction of an efficient conveyor load zone offering total material control is like the construction of a house: it starts with a good foundation. In a house the foundation consists of the footings and/or masonry walls of the basement; in a conveyor system the foundation is a stable, sag-free belt line.

For a conveyor to provide total control of dust and spillage, the design engineer should do whatever possible to keep the belt's travel line consistently steady and straight. If a belt were supported by a flat table that prevented its movement in any direction except in a straight line in the direction of carry, it would be easier to maintain a constant seal. Loaded or unloaded, stopped or in motion, the belt's position would be the same. The conveyor system designer must strive to achieve a support system for the belt that is similar to the table's ability to keep the running line true.

But this is not typically the way conveyors and their loading zones are designed. Many other factors–not the least of which is initial cost–are given higher priority than ensuring a stable belt line. And even when designed properly, transfer points are often so poorly maintained that the belt develops unwanted dynamic action that defeats the conveyor system's ability to supply total material control.

Among the factors that influence the running line of the conveyor are the loading of the belt with its cargo; the tensioning of the belt by the counterweight; the type, spacing, and arrangement of idlers; the construction and shape of the belt in the loading zone; the transition area; and the type of pulleys on the system.

Construction Basics

It is essential that the conveyor's support structure–the stringers–are straight; if they are not, they should be straightened or replaced. Laser surveying, discussed elsewhere in this volume, is the preferred method to verify stringer alignment.

Proper footings must be designed to provide a rigid support structure to prevent stringer deflection. The level of impact and amount of material being loaded must be considered to prevent excessive deflection under-load. Properly spaced stringer supports tied to rigid footings ensure a good base for the remaining structure.

Proper Belt Support

For an effective, minimum spillage transfer point, it is essential that the belt line be stabilized. If the belt is allowed to sag or flex under the stress of loading and high-speed belt movement, fines and larger pieces of material will work their way out, dropping onto the floor as spillage or becoming airborne as a cloud of dust. **(Figure 6.1)** Worse, these materials can wedge into entrapment points where they present the danger of gouging the belt or damaging other conveyor components.

A perfectly flat and straight belt line in the skirted area is essential to successfully sealing a transfer point. Any measurable belt sag–even a sag that is only barely apparent to the naked eye–is enough to permit fines to start a grinding action that leads to abrasive wear on the skirtboard seals and the belt surface. This wear allows the escape of materials into the environment. A groove cut into the belt cover along the entire belt **(Figure 6.2)** in the skirted area can almost always be attributed to material captured in entrapment points. These originate in belt sag or pinch points created in the construction of chute wall, wear liner, and skirting seal. In a properly designed transfer point, belt sag must be eliminated at all costs.

The key to a stable, sag-free line of travel is proper belt support. The crux of the problem arises in the difficulty in defining "proper." The amount of support needed is determined by the unique characteristics of each individual conveyor, transfer point, and material load. Included in the factors to be assessed are speed of travel, weight of material, loading impact, and troughing angle.

It is essential the belt be stabilized as far down the line as necessary. The support systems can extend beyond what is minimally required, with little harm other than an incidental increase in conveyor power requirements. But sometimes the belt support system is skimped on and left shorter than required. Then the fluctuations in belt travel at the end of the support system will create a spillage problem that will render the installed belt support system almost pointless. Belt support is like money; it is better to have a little extra than to fall a little short.

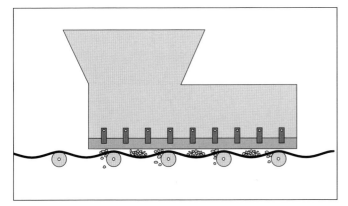

Figure 6.1
Belt sag under transfer points allows spillage and creates material entrapment points.

Figure 6.2
Material wedged into entrapment points in the conveyor structure can quickly abrade the belt.

Figure 6.3
Belt sag between idlers makes it difficult to seal the load zone.

Preventing Belt Sag

Belt sag is the vertical deflection of a belt from a straight line between idlers. It occurs more severely when there is insufficient tension on the belt and/or insufficient support below the belt to keep the belt taut. **(Figure 6.3)**

In Belt Conveyors for Bulk Materials, CEMA recommends that conveyor belt sag between idlers be limited to 2 percent for 35° idlers and 3 percent for 20° idlers. This reference notes, "Experience has shown that when a conveyor belt sags more than 3 percent of the span between idlers, load spillage often results."

This means under conventional four-foot (1.2 meter) idler spacing, the belt must be tensioned to limit sag to less than one inch (25 mm) on 35° idlers and less than 1.5 inches (38 mm) on 20° idlers. It should be noted these figures represent spacing not just under the transfer point but along the full length of the conveyor. It is particularly important to control sag in the transfer point where the load undergoes changes in height, speed, and direction. These alterations induce forces into the load that can carry fines and dust out of the system or push these particles into pinch points if belt sag creates the opportunity.

One method that can reduce (but not eliminate) sag from the entire length of the conveyor is to increase the belt tension. However, there are drawbacks to this, including increased drive power consumption and additional stress on belting and belt splices. The real benefit of proper tension is to prevent slippage, rather than preventing sag in the loading zone.

To prevent spillage and contain fine dust particles as required by today's regulatory standards, necessitates the elimination of belt sag. After achieving proper belt tension to suit belt, conveyor, and system loading, the method of preventing belt sag is improvements in the belt's support systems. To prevent fugitive material in a loading zone, it is important that belt sag be controlled to a maximum of one-half of one percent (0.005). That is an allowable sag of approximately 1/4 inch (6 mm) in a distance of four feet (1.2 m).

There are a number of components that can be used–independently or in combination–to control belt sag by improving belt support in the loading zone.

Idlers

Idlers are the most numerous of conveyor components, both in terms of the number used in a particular conveyor and in the number of styles and choices available. There are many types, but they all share the same responsibilities: they shape and support the belt, and they minimize the power needed to transport the materials.

As the belt moves through its cycle of loading, carrying, discharge, and return, it must be supported to prevent spillage, maintain tracking, and prevent damage. In addition, the idlers provide the otherwise

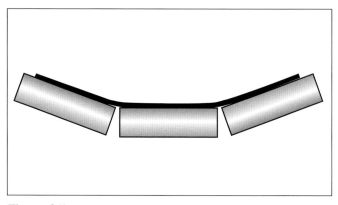

Figure 6.5
Standard troughed idler sets incorporate three rollers of the same length.

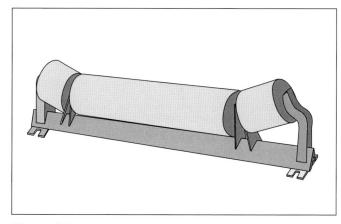

Figure 6.6
Picking idlers use short rollers on each end of a longer center roller.

IDLER CLASSIFICATION			
Classification	Former series no.	Roll diameter (in.)	Description
B4	II	4	Light Duty
B5	II	5	
C4	III	4	Medium Duty
C5	III	5	
C6	IV	6	
D5	NA	5	
D6	NA	6	
E6	V	6	Heavy Duty
E7	V	7	

Figure **6.4**
Idler Classifications as established by CEMA.

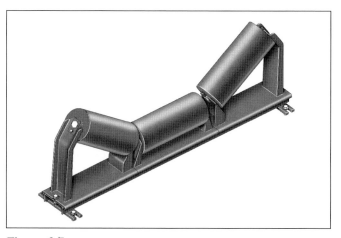

Figure 6.7
Inline Troughed Idler.

shapeless belt with the required form. In the loading and carrying sections, troughing idlers shape the flat belt into a channel to ensure centered loading and minimize spillage.

Idlers are classified or rated based on roll diameter, type of service, operating condition, belt load, and belt speed. CEMA uses a two-character code that expresses both idler diameter in inches and the load rating with rankings from B4 to E7. **(Figure 6.4)**

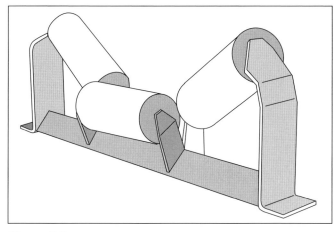

Figure 6.8
Offset Troughed Idler.

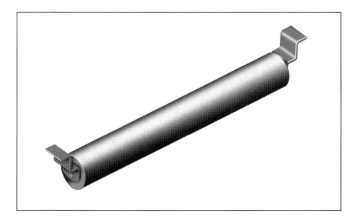

Figure 6.9
Flat Return Idler.

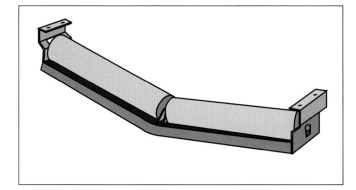

Figure 6.10
V-Return Idler.

Whatever its rating, an idler consists of one or more rollers–usually containing one or more bearings to ensure it is free-rolling–supported or suspended by a framework that is installed across the conveyor stringers.

Carrying Idlers

Carrying idlers are available in flat or troughed designs. The flat design consists of a single horizontal roll for use with flat belts.

The troughed idler usually consists of three rolls–one flat roll in the center with an inclined roll on either side of it. **(Figure 6.5)** Typically, these rolls are the same length, although there are sets that incorporate a longer center roll and shorter inclined "picking" idlers. This design supplies a larger flat area to carry material while allowing inspection or "picking" of the cargo. **(Figure 6.6)**

Troughed idler sets are available as in line idlers **(Figure 6.7)**, where the centerlines of the three rolls are aligned) and offset idlers **(Figure 6.8)** (where the centerline of the center roll is alongside the plane of the wing idlers). This reduces the overall height of the idler set and, accordingly, is popular in underground mining applications where headroom is at a premium. Offset idlers eliminate the gap between the rollers, reducing the chance of junction joint belt damage.

Return Idlers

Return idlers provide support for the belt when it is on its way back to the loading zone after unloading its cargo. These idlers normally consist of a single horizontal roll mounted to the underside of the conveyor stringers. **(Figure 6.9)** V-return idlers incorporating two rolls are sometimes used to reduce energy requirements and improve belt tracking. **(Figure 6.10)**

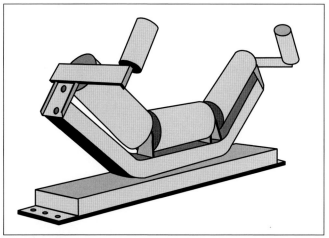

Figure 6.11
Carrying Side Training Idler.

Return Idler Spacing	Belt Width	Carrying Side Idler Spacing Outside the Load Zone					
		Weight of Material Handled in lb per ft³ (kg/m³)					
		30 (480)	50 (800)	75 (1200)	100 (1600)	150 (2400)	200 (3200)
ft (m)	in. (mm)	ft (m)	ft (m)	ft (m)	ft (m)	ft (m)	ft (m)
10.0 (3)	18 (457)	5.5 (1.7)	5.0 (1.5)	5.0 (1.5)	5.0 (1.5)	4.5 (1.4)	4.5 (1.4)
10.0 (3)	24 (610)	5.0 (1.5)	4.5 (1.4)	4.5 (1.4)	4.0 (1.2)	4.0 (1.2)	4.0 (1.2)
10.0 (3)	30 (762)	5.0 (1.5)	4.5 (1.4)	4.5 (1.4)	4.0 (1.2)	4.0 (1.2)	4.0 (1.2)
10.0 (3)	36 (914)	5.0 (1.5)	4.5 (1.4)	4.0 (1.2)	4.0 (1.2)	3.5 (1.1)	3.5 (1.1)
10.0 (3)	42 (1067)	4.5 (1.4)	4.5 (1.4)	4.0 (1.2)	3.5 (1.1)	3.0 (0.9)	3.0 (0.9)
10.0 (3)	48 (1219)	4.5 (1.4)	4.0 (1.2)	4.0 (1.2)	3.5 (1.1)	3.0 (0.9)	3.0 (0.9)
10.0 (3)	54 (1372)	4.5 (1.4)	4.0 (1.2)	3.5 (1.1)	3.5 (1.1)	3.0 (0.9)	3.0 (0.9)
10.0 (3)	60 (1524)	4.0 (1.2)	4.0 (1.2)	3.5 (1.1)	3.0 (0.9)	3.0 (0.9)	3.0 (0.9)
8.0 (2.4)	72 (1829)	4.0 (1.2)	3.5 (1.1)	3.5 (1.1)	3.0 (0.9)	2.5 (0.8)	2.5 (0.8)
8.0 (2.4)	84 (2134)	3.5 (1.1)	3.5 (1.1)	3.0 (0.9)	2.5 (0.8)	2.5 (0.8)	2.0 (0.6)
8.0 (2.4)	96 (2438)	3.5 (1.1)	3.5 (1.1)	3.0 (0.9)	2.5 (0.8)	2.0 (0.6)	2.0 (0.6)

Metric measurements are conversions from CEMA measurements.

Figure 6.12
Recommended Idler Spacing as established by CEMA.

Figure 6.13
On this conveyor, idler spacing has been reduced to improve belt support.

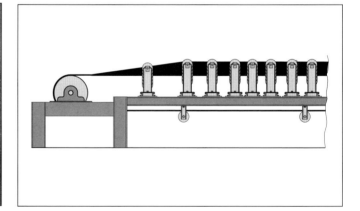

Figure 6.14
Close idler spacing is used to eliminate sag and increase support of the belt.

Training Idlers

There are a number of designs for training idlers that work to correct the belt path. **(Figure 6.11)** Typically, these idlers are self-aligning; they react to the movement of the belt to move to a position that will steer the belt back into the center. They are available for both carrying side and return side application. *For more information, see Chapter 15 Tracking.*

Idler Spacing

The spacing between the rolling components has a dramatic effect on both the support and shape mission of the idlers. Idlers placed too far apart will not properly support the belt or enable it to maintain the desired profile. Placing idlers too close together will improve support and belt profile, but will increase conveyor construction costs and power requirements.

Normally, idlers are placed close enough to support the belt so it does not sag excessively between them. If there is too much sag, the load will shift as the belt carries it up over each idler and down into the valley between. This shifting of the load will increase belt wear and power consumption. A typical specification calls for a limit of three percent sag. In fact, this is too much sag for the loading zone. Here, where sag allows material entrapment and spillage, sag should be limited to one-half of one percent. **Figure 6.12** presents CEMA's table of recommended idler spacing outside the load zone with metric conversions added.

The spacing of return idlers is determined by belt weight, since no other load is supported by these idlers, and sag-related spillage is not a problem on this side.

Idlers in the Skirted Area

The basic and traditional way to improve belt support is to increase the number of idlers under the loading zone. **(Figure 6.13)** By increasing the number of idlers in a given space–and consequently decreasing the space between the idlers–the potential for belt sag is reduced. Idlers can usually be positioned so that their rolls are within one inch (25 mm) of each other. **(Figure 6.14)**

However, this solution is not without a complication: as the idlers are bunched more closely together, it becomes more difficult to service them.

Idlers are typically maintained by laying the framework over on its side parallel to the belt. If the idlers are closely spaced, there is no room available for service. **(Figure 6.15)** To reach one set of idlers, one or more adjacent sets must be removed, creating a "falling domino" chain reaction.

Idlers that slide into position avoid this problem. **(Figure 6.16)** In this case, the rollers are mounted on a frame or track that allows straight-in, straight-out installation perpendicular to the belt's path. This means each individual set can be serviced without laying the frames on their sides or raising the belt. Typically these systems are designed to allow service from either side. The track upon which idlers (and/or other belt support components) slide can be added in as a supplement to

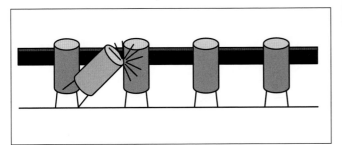

Figure 6.15
Close idler spacing may create service problems.

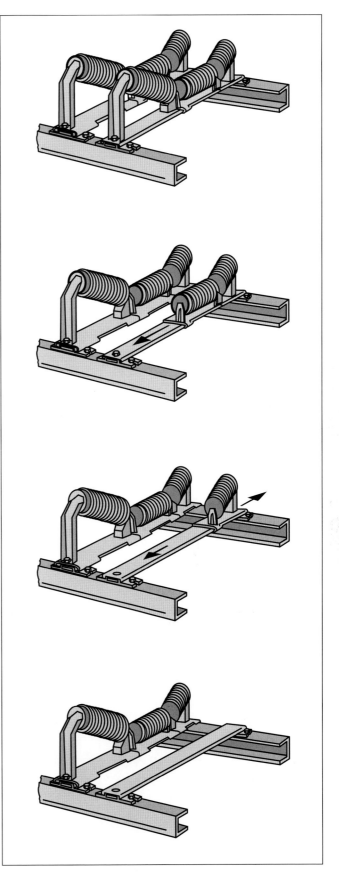

Figure 6.16
To make service easier, idlers can be modular and mounted on tracks.

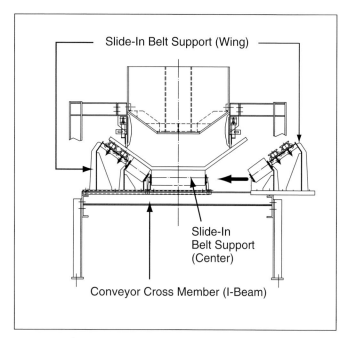

Figure 6.17
Belt support structures can be designed to slide on the conveyor cross beams.

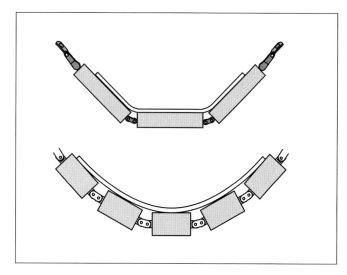

Figure 6.18
3-Roll and 5-Roll Catenary Idlers.

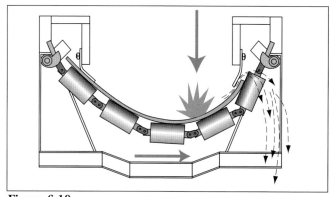

Figure 6.19
Impact loading on catenary idlers can cause spillage over the edge of the belt.

the conveyor structure, or it can be incorporated as part of the structure using the I-beam cross braces under the loading zone. **(Figure 6.17)** Incorporating this slide-in-place system in the conveyor's design stage allows the use of modular belt support structures–idlers, cradles, or combination units–and simplifies component installation. This is particularly beneficial on wide belts where large components may require cranes or other heavy equipment for installation.

Idlers must be aligned with care and matched so as not to produce humps or valleys in the belt. Idlers should be checked for concentricity; the more they are out of round, the greater tendency for the belt to flap or bounce. Only idlers supplied by the same manufacturer and of the same roll diameter, class, and trough angle should be used in the skirted area of a conveyor. Even a slight difference in an idler's dimensions–the difference from one manufacturer to another–can create highs and lows in the belt line, making it impossible to provide effective sealing. Never install troughed belt training idlers within the skirted area, as they sit higher than the adjacent regular carrying idlers and raise the belt as they swivel.

When installing idlers in a transfer point, they should not only be square with the stringers but also aligned horizontally and vertically across the conveyor. If idlers are not correctly aligned, they create highs and lows in the belt line. These variations will cause pinch points, entrapping material that will lead to belt damage and spillage. Laser surveying can be used to ensure the alignment of all rolling components.

Catenary Idlers

Catenary idlers are sets of rollers–typically three or five rolls–linked together on a cable, chain, or other flexible connections, and suspended on a chain or flexible mount below a conveyor loading zone. **(Figure 6.18)** This mounting allows the idler to be quickly moved or serviced and provides some amount of conveyor and load centering.

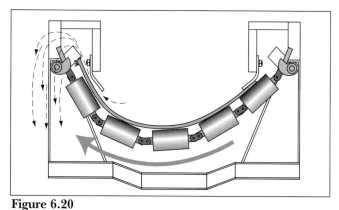

Figure 6.20
The side-to-side movement allowed by catenary idlers can lead belt damage.

These idler sets swing freely under the forces of loading material acting to absorb impact and centralize the load. Catenary idlers are typically seen in very heavy-duty applications; that is, conveyors seeing high impact levels and large volumes of material. Typical installations would include at the discharge of bucket wheel excavators and under the loading zone of long overland conveyors carrying run-of-mine material. They are also commonly used in the foundry industry.

However, the "bounce" and changes in belt path catenary idlers add to the conveyor must be considered when engineering the conveyor system. Because of their design intention to swing and reshape the belt, catenary idlers **(Figure 6.19)** can create problems in maintaining an effective seal. As it swings, the catenary set moves the belt from side to side. This allows the escape of fugitive material out the sides of the loading zone and creates mistracking that exposes belt edges to damage from the conveyor structure. **(Figure 6.20)**

Catenary Idler Stabilizers

This flaw in the function of catenary idlers has led to the development of a hybrid system to stabilize the catenary system. **(Figure 6.21)** This catenary idler stabilizer maintains the load re-shaping and impact absorption of the catenary idlers, while controlling the swinging motion. This minimizes load spillage and controls mistracking. **(Figure 6.22)**

Available in designs compatible with three-or five-roll catenary idlers, the stabilizer system replaces the upper rolls on adjoining sets of idlers with urethane blocks. The system's slider blocks are installed in place of the upper rollers on either side of the roller set to maintain a stable and sealable belt edge.

Heavy-duty, energy-absorbing shock cords connect the suspended idlers to the support system to cushion the impact and prevent the irregular movements that create spillage and mistracking. **(Figure 6.23)**

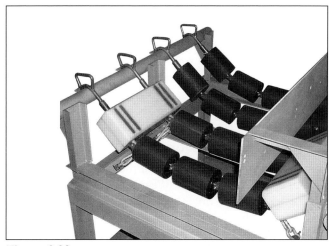

Figure 6.22

A catenary idler stabilizer (foreground) assumes the same profile as the catenary idlers (at the rear).

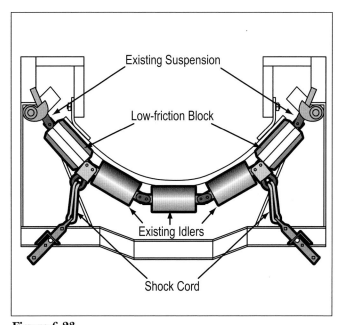

Figure 6.23

The shock cords absorb the impact of loading while low-friction blocks hold the belt edge steady.

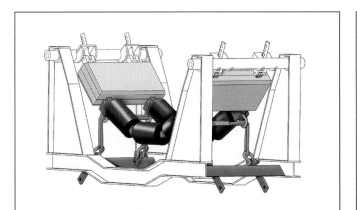

Figure 6.21

A catenary idler stabilizer will control the movement of the suspended idler to improve sealing.

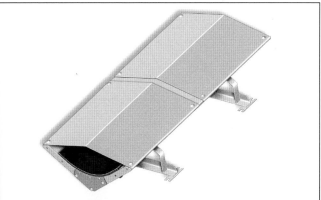

Figure 6.24

Air-supported belt conveyors replace carrying side idlers with air.

Belt Support

The catenary idler stabilizer system combines the extreme impact absorption and quick-release capabilities of catenary idlers with the improved belt sealing and tracking abilities of belt support cradles. Originally designed as a retrofit solution to upgrade existing catenary idler installations, the stabilizer system is now also available for new conveyor systems.

Air Supported Conveyors

Another concept for stabilizing the belt path is the air-supported belt conveyor. **(Figure 6.24)** As discussed in Chapter 1, these conveyors replace carrying side idlers and cradles with a trough below the belt. **(Figure 6.25)** A series of holes in the center of the trough allow air supplied by a low-power fan to raise the belt. This allows the belt to travel with less friction and lower energy consumption than comparable idler conveyors.

Because there are no idlers, the air-supported belt conveyor maintains a stable line of travel. There is no belt sag, which reduces spillage. The 1/32 inch (0.8 mm) film of air rises from the trough-shaped plenum to uniformly support the belt and create a flat belt line, eliminating the need for an engineered sealing system. **(Figure 6.26)**

Performance of an air conveyor depends on how successful the conveyor's feeding system or chutes are in providing a consistent and central material load.

Air-supported belt conveyors are designed for light-duty applications–typically materials that are two–inches in diameter (50 mm) or smaller. They should not be subjected to impact loading beyond the CEMA light-duty impact rating.

New designs allow for the conversion of existing troughed idler conveyors, or even parts of existing conveyors, to air–supported belt conveyor systems. **(Figure 6.27)** *For additional material on air-supported belt conveyors, see the Alternative Conveying Systems section in Chapter 1: Belt Conveyors.*

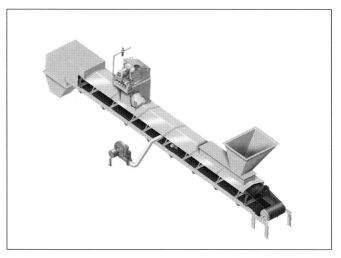

Figure 6.26
The air-supported conveyor maintains a stable belt profile without sag.

Figure 6.27
Air supported conveyors can be installed to upgrade portions of conventional idler conveyors.

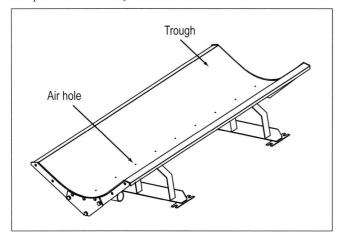

Figure 6.25
Air from a fan is released through a series of holes in the trough, reducing friction and power requirements.

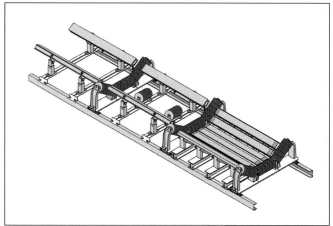

Figure 6.28
Slider beds or cradles are now often used in place of idlers to support the belt.

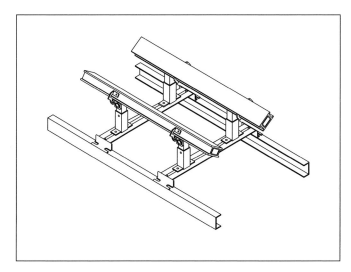

Figure 6.29
"Side rail" belt support cradles maintain a stable belt profile.

Figure 6.30
Side rail cradles support the belt edges to improve belt sealing.

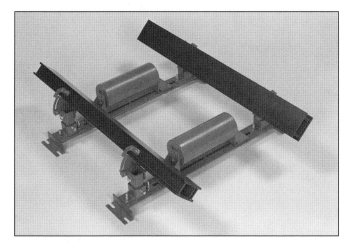

Figure 6.31
An optional center roller can be added to improve belt support.

Using Cradles for Belt Support

So important is the flat table concept to good sealing that many designers now use slider beds or cradles in place of idlers under transfer points. **(Figure 6.28)** Instead of using the rolling "cans" of idlers to support the belt, cradles use some sort of bar. Typically supplied in four-foot (1.2-meter) lengths, these bars provide a smooth, even surface that allows the belt to slide across on a steady, reduced-friction path.

Used with steel roll idlers or impact idlers at their entry and exit, these cradles provide a smooth, low-friction, fixed-in-place surface on which the belt can move, and against which an effective seal can be formed to prevent the escape of fugitive material.

In addition to improved sealing, other benefits of the use of cradles under the transfer point include a reduction in moving parts and required lubrication. The modular design of the cradles allows the belt support system to be extended as far as the circumstances require.

Bar support systems are of limited life in high-load, very high-speed applications. As noted above, catenary impact idlers are often used in these situations.

Seal Support Cradles

One form of cradle consists of slider bars in a "side rail" configuration. **(Figure 6.29)** This system places one or more low-friction bars on either side of the conveyor directly under the chutework. The bars function like the rails of a bed, holding up the sides of the belt to allow effective sealing of the belt edge. **(Figure 6.30)**

It must be noted there is a difference between edge sealing bars that are designed to provide continuous support at the belt edges only, and impact bars that must absorb loading impact underneath the belt (in addition to providing sealing support). Impact cradles are discussed at greater length later in this section.

Each cradle installation may be one or more cradles long, depending on the length of the loading zone, the speed of the belt, and other conveyor characteristics. These support cradles are installed in line with the entry and exit idlers (as well as with the intermediate idlers placed between the cradles) to avoid the creation of entrapment points.

In some circumstances where full width support is required and there is ample conveyor power, the side rail slider bar configuration should be supplemented by an additional support in the middle of the belt. This support can be provided with an additional "third" rail or with center idler rollers. **(Figure 6.31)**

On faster, wider, more heavily loaded belts, the cradles may also need additional bars on each side. **(Figure 6.32)**

Edge support slider bars can be manufactured from ultra-high molecular weight (UHMW) polyethylene or other low-friction plastic. They provide a low-drag, self-lubricating surface that allows the belt to glide over it without heat accumulation or undue wear on either belt or bar. One system is designed so the bars are formed in an "H" or "box" configuration, which allows use of both the top and bottom surfaces. **(Figure 6.33)**

At conveyor speeds above 750 fpm (3.8 m/sec), the heat created by the friction of the belt on the bars can reduce the performance of the plastic bars. Consequently, the use of stainless steel support bars has found acceptance in these applications. The stainless steel bars should also be incorporated in applications with service temperatures over 180° F (82° C). Some countries require anti-static, fire–resistant materials (commonly called FRAS) be used for underground conveyor components in contact with the belt; other nations require fire resistant ratings equivalent to MSHA requirements.

The bars should be supported in a mounting frame that is adjustable to allow easy installation, alignment, and maintenance. This adjustable frame will accommodate various idler combinations and chute wall widths.

Design features that should be included in the engineering of these sealing support cradles include a method for vertical adjustment of the bar to improve ease of installation and allow compensation for wear. The support bars should be held in support positions without the risk of fasteners coming into contact with the belt; i.e., the bolts should be installed parallel rather than perpendicular to the belt.

An edge support cradle made with these bars may add incrementally to the friction of the conveyor belt and to the conveyor's power requirements. But this marginal increase in energy consumption is more than offset by the elimination of the expenses for the cleanup of skirt leakage and the unexpected downtime necessary for idler maintenance or belt replacement.

Impact-Absorbing Systems

Nothing can damage a conveyor's belting and transfer point structure and create material leakage as rapidly and dramatically as loading zone impact from heavy objects or sharpened edges. **(Figure 6.34)** Whether arising from long material drops or large lumps, whether boulders, timber, or scrap metal, these impacts can damage components like idlers and sealing strips. Impact can create a "ripple" effect on the belt, de-stabilizing its line of travel and increasing the spillage of material. Heavy and repeated impact can also damage the belt cover and weaken its carcass.

To control impact, conveyor transfer points should be designed to minimize the height of material drop and to

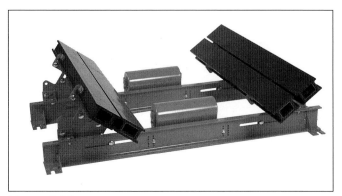

Figure 6.32

For more challenging applications, the cradle may be equipped with additional bars.

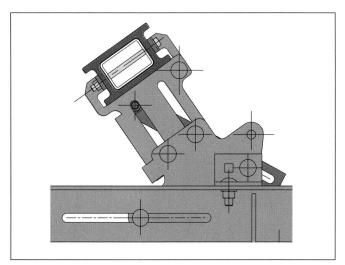

Figure 6.33

Support bars designed in a "box" or "H" shape provide a double wear life.

Figure 6.34

Impact in the loading zone can originate in large lumps or long material drops.

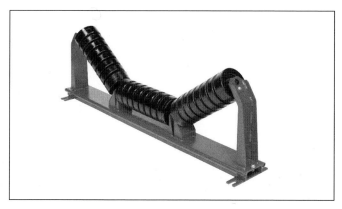

Figure 6.35

Impact idlers are typically composed of resilient rubber disks.

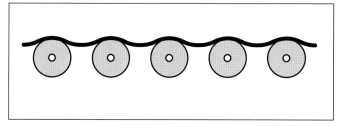

Figure 6.36

Impact idlers can allow belt sag between rollers.

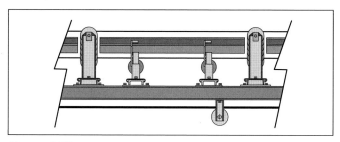

Figure 6.37

Bar support systems provide support over their full length to prevent sag.

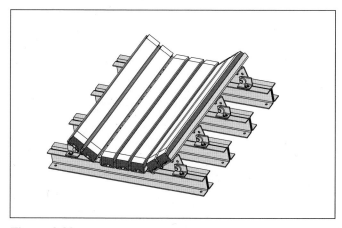

Figure 6.38

Impact-absorbing cradles are commonly used in conveyor loading zones.

use a low-angle trajectory so lumps strike the belt at a grazing angle. Other systems to control impact can also be employed. When the material consists of a mixture of fines and lumps, the loading chute can be arranged to deposit the fines first and then the lumps onto the belt being loaded. This design uses the material fines to form a cushion of sorts, which absorbs the impact and shields the belt cover from the blows of the larger lumps. This "cushion" can be formed through the use of a grizzly or chute screen or a wedge-shaped loading chute, where fines can fall through the first portion of the opening, but larger pieces move farther down the line before dropping onto the belt.

The use of engineered "hood and spoons" chutes that control the acceleration of the material and create a shallow loading trajectory is increasing. These chutes are more expensive to engineer, fabricate, and install, but they can greatly reduce impact, dust, and pressure on the sealing system.

In most cases it is not possible to totally eliminate impact, so it will be necessary to install some sort of energy-absorbing system under the loading zone.

Impact Idlers

Rubber-cushioned impact idlers are one solution for absorbing impact in the belt's loading zone. **(Figure 6.35)** Idlers with rollers composed of resilient disks are installed under the belt in the impact zone to cushion the force of loading.

But the problem encountered with impact rollers is this: each idler only supports the belt at the top of each roller. **(Figure 6.36)** No matter how closely they are spaced, the rounded shape of the idler allows the conveyor belt to oscillate or sag away from the ideal flat profile. This sag allows and encourages the release or entrapment of fugitive material, which grinds away at components like skirtboard seals, idler bearings, and even the idler rollers themselves to further impair conveyor performance. The spacing between impact rollers offers little protection from tramp materials penetrating the belt.

In addition, impact idlers with rubber disc rollers are more likely to be out of round than idlers with conventional "steel can" rollers. Even when they are tightly spaced together, there is likely to be some gap (and hence sag) between the rollers.

Even impact idlers are subject to impact damage. Idlers with worn or seized bearings can cause the belt to run erratically, allowing mistracking and spillage over the sides of the belt. Idlers that have been damaged from severe impact or that have seized due to fugitive material will increase the conveyor's power consumption significantly.

The use of a bar system provides full-length support, with minimal gaps in the support system. **(Figure 6.37)**

In many cases it becomes more effective to seal with an impact cradle.

Impact Cradles

A bar or pad support system provides continuous support along its full length, making the use of four-foot-long (1.2 m) loading zone cradles composed of impact-absorbing bars a common practice. **(Figure 6.38)** Impact cradles are installed directly under the impact zone to bear the brunt of the shock of the material loading. **(Figure 6.39)** These cradles are usually composed of individual impact bars assembled into a steel support framework that is then installed under the loading zone. **(Figure 6.40)** The bars are composed of durable elastomeric materials that combine a slick top surface, allowing the belt to skim over it to minimize friction and a sponge-like secondary layer to absorb the energy of impact.

Some manufacturers align the bars parallel with the direction of belt travel; others use modular segments that align to form a saddle that is perpendicular to belt travel. Some use a bar that features a slick top surface and a cushioned lower layer permanently attached; others feature separate components that are put together at the application.

The impact cradle is usually installed so that the bars in the center of the cradle are set slightly–0.5 to 1 inch (12 to 25 mm)–below the normal unloaded track of the belt **(Figure 6.41)**. This allows the belt to absorb some of the force of impact when the material loading deflects it down onto the cradle, while avoiding continuous friction and wear when the belt is running empty. The wing bars–the bars on the sides of the cradle–are installed in line with the entry, exit, and intermediate idlers to prevent belt sag and pinch points.

As many impact cradles as necessary should be installed to support the belt throughout the impact zone. It is generally recommended that idlers be installed immediately before and after each impact cradle to provide efficient carriage of the belt.

In some impact areas it may be useful to go up as far as eight feet (2.4 m) between idlers. These applications might include long loading zones where it is difficult to predict the location of the impact and where rollers might be damaged by point impact loading. These would also include transfer points under quarry and mine dump hoppers, pulp and paper mills where logs are dropped onto belts, or recycling facilities that see heavy objects from car batteries to truck engines dropped on conveyors.

The impact absorbing bars installed in the cradles are generally composite bars formed of a top layer of a slick, low-friction material like UHMW polyethylene with an energy-absorbing layer of rubber underneath. **(Figure 6.42)** Other materials have been tried, including

Figure 6.39
Installation of impact cradles can prevent damage to the belt and conveyor structure.

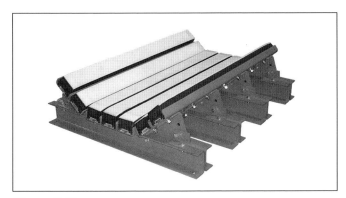

Figure 6.40
Impact cradles are typically composed of a number of energy-absorbing bars held in a steel framework.

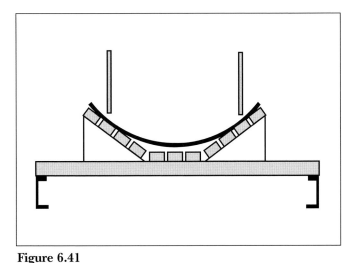

Figure 6.41
The center bars of an impact cradle are installed below the belt's unloaded travel line, while the wing bars are in line with the wing idlers.

top covers of fire-retardant, MSHA-accepted polyurethane, and nylon, and polyurethane in a honeycomb pattern as an absorption layer. These bars are mounted and supported in a heavy-duty steel base. Installation of impact cradles is simplified through the use of adjustable wing supports, which simplify installation by allowing the cradle to be slid under the belt in a flat form, and then the sides raised to the appropriate trough angle. The wing supports are also adjustable to allow for variations in idler construction. The cradle should incorporate some mechanism to adjust bar height to accommodate wear. **(Figure 6.43)**

Again, note the difference between impact-absorbing bars and bars designed to provide only continuous support. Impact bars could be installed in a "side rail" frame to provide edge support; however, their requirement for impact-absorption tends to make this construction more expensive than using bars designed expressly for support.

Impact cradles are available in a welded construction or models that feature sub-assemblies held in place with bolts or pins. The bolt and pin versions offer the advantages of easier installation and maintenance. An additional step toward improved maintenance is the use of a track-mounted design, which allows faster replacement of bars when required.

A Standard for Impact Cradles

The Conveyor Equipment Manufacturers Association published in CEMA Standard 575-2000 in 2000 an engineering and selection standard for impact bed/cradle design. **(Figure 6.44)**

Like many CEMA standards, CEMA Standard 575-2000 was developed to give manufacturers and users a common rating system to reduce the chance of misapplication. This standard provides a method for determining the proper rating of impact cradles for various applications.

The benefit of this standard is a user-friendly classification system. With this system users can quickly establish the duty rating and dimensional standards for their application. The classification system is based on the impact energy created by the bulk solid, and the dimensions are established from the CEMA idler classification. The ratings are Light, Medium and Heavy Duty, as indicated by initial letters L, M, and H respectively.

To determine the duty rating, a user need only multiply the weight of the largest lump (W) by the maximum fall height (h). This yields a reference number for impact force, which is then used to select one of the three ratings from a simple chart. For example, a 25-pound block falling 15 feet equals 375 lb.-ft of force. A rough equivalent using metric measurements is a 10-kg lump falling 5 meters, resulting in 50 kg-m of impact energy. As seen in Figure 6.44.,

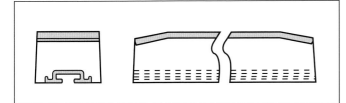

Figure 6.42
Impact–absorbing bars are commonly composed of two layers: a low friction layer on top and a cushion layer on the bottom.

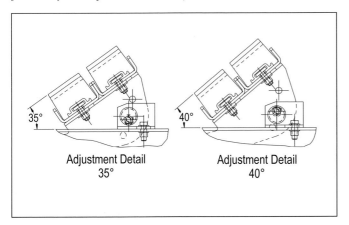

Figure 6.43
The impact cradle should allow for adjustment of bar height to compensate for wear.

Code	Rating	Impact Force lbf.	Impact Force kN	W x h (ref.) lb.-ft.	W x h (ref.) kg-m
L	Light Duty	< 8,500	<37.8	< 200	<28
M	Medium Duty	8,500 to 12,000	37.9 to 53.4	200 to 1000	28.1 to 138
H	Heavy Duty	12,000 to 17,000	53.5 to 75.6	1000 to 2000	138.1 to 277

Figure 6.44
CEMA 575-2000 Impact Bed/Cradle Ratings (Metric measurements are converted from imperial measurements and added for reference).

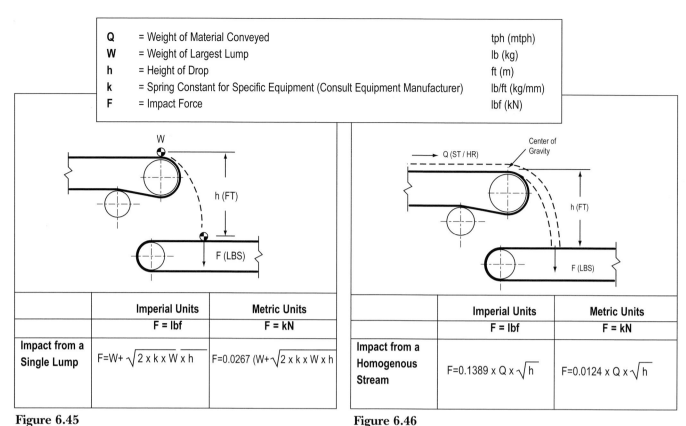

Figure 6.45

CEMA calculation of impact force from a single lump. (Metric measurements are added for reference).

Figure 6.46

CEMA calculation of impact force from a homogenous stream of material. (Metric measurements are added for reference).

both of these would be considered medium-duty applications. The dimensional standard is the same as CEMA idler ratings: B, C, D, E, or F, followed by the nominal idler diameter as measured in inches (i.e., 5, 6, and 7). If in this case the conveyor was using idlers with a six-inch-diameter roller and a "C" rating, the user would specify an M-C6 Impact Bed for this application.

A second portion of the CEMA project was to develop a standard method for calculating the impact capacity of the products of individual manufacturers. The CEMA task force carried out a great deal of comparison between the simplified method in the classification system and performance of specific designs. Calculations were done, including the belt, the impact bar, the impact bed structure, and the conveyor structure. Data such as the spring constant of the belt and the impact bars is not available in many cases, and when it is, it may be considered proprietary information by the manufacturers and difficult for the user to obtain for calculation purposes.

Following the method of the CEMA standard, the impact force is determined by calculating the worst-case impact. For a given application, both the impact from a single lump **(Figure 6.45)** and the continuous flow **(Figure 6.46)** should be calculated. The equations used by CEMA are generally accepted as reasonable approximations of impact forces.

The impact from a single lump almost always yields the highest impact force and, therefore, governs the impact rating that should be specified.

Based on the results of these comparisons, the CEMA work group determined that the simplified Light-Medium-Heavy classification system cited above provides simplicity of use with acceptable accuracy. Additionally, the committee agreed a structural design "safety" factor of 1.5 would provide reasonable room for error and allowance for conditions outside normal estimates.

The CEMA Standard 575-2000 Impact Beds/Cradles provides a conservative and easy-to-use rating system for impact cradles used in bulk handling applications. While there are many assumptions in the CEMA standard, the availability to the user of a simple reference calculation, (W x h), is justified and very useful.

These calculations provide a conservative method for evaluating impact to allow users to apply it to any material handling application within its scope and achieve an acceptable response. By following the dimensional guidelines and using the sample specification in the standard, the misapplication of impact cradles will be greatly reduced.

Cradles With Bars and Rollers

Hybrid designs are popular as a way of combining the low power requirements of rollers and the flat sealing surface of the impact bar. A number of "Combination Cradle" designs are available, which use bars for a continuous seal at the belt edge, but incorporate rollers under the center of the belt. (**Figure 6.47**) These designs are most common on high-speed conveyors operating above 750 fpm (3.8 /sec) or applications where there is a heavy material load that would create high levels of friction or drag in the center of the conveyor.

With these designs, the belt center is supported on conventional parallel rollers. Since this is where the majority of the material load is carried, running friction is kept low, reducing the power consumption of the conveyor. Since the belt is continuously supported, belt deflection between idlers is eliminated, and as a result, spillage in this area is reduced to a minimum. Since the central rollers operate in a virtually dust-free environment, bearing and seal life is significantly extended, thus reducing long-term maintenance costs.

It is also possible to design cradles incorporating impact bars in the center with short picking idlers closely spaced on the wings. Here the design intention is to provide superior impact cushioning in the center of the belt, while reducing friction on the belt edges.

For loading zones with the highest levels of impact, the entire impact cradle installation can be cushioned. By mounting the cradle on a shock-absorbing structure, the cradle can absorb even more impact. (**Figure 6.48**) While this does reduce the stiffness of the entire loading zone and absorb impact force, it has the drawback of allowing some vertical deflection of the belt in the skirted area, and so makes it harder to seal the transfer point.

Loading in the Transition Area

Generally speaking, the belt should be fully troughed before it is loaded; there should be no loading in the conveyor's transition area. But in underground mining and other applications where space is at a premium, conveyor designers need to load the belt as close to the tail pulley as possible. This can create problems such as spillage, belt damage, and belt mistracking. The practice of loading in the transition area should not be incorporated into new conveyor designs because the problems this practice creates are much more costly than the one-time capital savings it provides.

To minimize damage and spillage in the transition area, impact cradles have been constructed that support the belt as it changes from flat to a 20° trough. (**Figure 6.49**) This "transition cradle" allows loading in the conveyor's transition area. However, this assembly

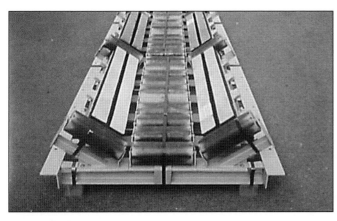

Figure 6.47
Combination cradles incorporating both bars and rollers are useful for some applications.

Figure 6.48
Installing a cradle on shock-absorbing air spring allows it to absorb higher levels of impact.

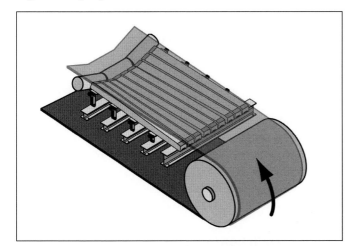

Figure 6.49
A "transition cradle" changes from flat to a 20° trough to allow loading in the conveyor's transition area.

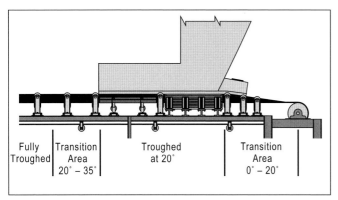

Figure 6.50
Two-stage troughing may be used where the transition area is too short for the belt to be fully troughed before loading.

Figure 6.51
Lengthy transfer points may require the use of both impact and seal support cradles.

Figure 6.52
Intermediate idlers are placed between bar support structures to reduce conveyor drag.

requires custom engineering to be precisely matched to the dimensions of the conveyor and pulley. This can make the cost of this cradle prohibitively high.

Another solution where transition distance is too short to allow full toughing before the loading zone is two-stage troughing. **(Figure 6.50)** In this case the belt is troughed to 20° before it enters the loading area. After the skirted area, when the load has settled, the belt undergoes another transition to the 35°-angle trough used for the length of the conveyor.

Multiple Cradle Systems

In the interest of stabilizing the belt line until the load settles on the conveyor, it is often appropriate to install combination systems. **(Figure 6.51)** These employ one or more impact cradles under the actual loading zone, followed by a number of "side rail" support cradles that support the belt until the load has settled. These combination systems provide an efficient way to combine optimum belt support with maximum cost efficiency.

The distance required to stabilize the load on the belt indicates the number of cradles required. Common "ball park" figures typically estimate two to three feet of cradle per 100 feet-per-minute (.66 to 1 meter per 0.5 m/sec) of belt speed.

Intermediate Idlers

Some designs call for cradles to be installed "back-to-back" for extended distances. However, the use of intermediate idlers between cradles is a sound engineering idea. **(Figure 6.52)** Installing idlers between two cradles (or putting each cradle between two idlers) will reduce the drag of the conveyor belt over the bars. This reduces the conveyor's amp draw, and as a result, increases its load-carrying capacity. It also minimizes friction and power consumption when the belt is moving but unloaded. In addition, the heat build-up in the bars will be reduced, and the belt support cradles will have a longer life expectancy.

Idlers should be specified before and after each four-foot (1200-mm) cradle; the number of idler sets required is the same as the number of cradles required plus one. To ensure uniformity for a stable belt line, all of these idlers should be of the same brand with the same size roller.

Idlers should be of the same type as would be installed if there were no cradles. Impact idlers should be used between cradles under the loading zone; conventional idlers can be used outside the impact area.

Cradle Installation

Cradles can be welded or bolted to the stringers; it may be better to bolt the system in place, as this will

Belt Support 6

Average Belt Weight lbs/ft (kg/m)			
Belt Width in. (mm)	Material Load On Belt [as measured in lbs/ft³ (kg/m³)]		
	30-74 (500-1199)	75-129 (1200-2066)	130-200 (2067-3200)
18 (400-500)	3.5 (5.2)	4 (6.0)	4.5 (6.7)
24 (500-650)	4.5 (6.7)	5.5 (8.2)	6 (8.9)
30 (650-800)	6 (8.9)	7 (10.4)	8 (11.9)
36 (800-1000)	9 (13.4)	10 (14.8)	12 (17.9)
42 (1000-1200)	11 (16.4)	12 (17.9)	14 (20.8)
48 (1200-1400)	14 (20.8)	15 (22.3)	17 (25.3)
54 (1400-1600)	16 (23.8)	17 (25.3)	19 (28.3)
60 (1600-1800)	18 (26.8)	20 (29.8)	22 (32.7)
72 (1000-2000)	21 (31.3)	24 (35.7)	26 (38.7)
84 (2000-2200)	25 (37.2)	30 (44.6)	33 (49.1)
96 (2200-2400)	30 (44.6)	35 (52.1)	38 (56.6)

Figure 6.53

Average Weight of Unloaded Belting, per CEMA. (Metric measurements are converted from imperial measurements and added for reference).

allow more efficient maintenance. It is important the cradle be engineered to allow some simple means of adjustment. This will enable the cradle to work with idlers of varying manufacture and allow compensation for wear.

The wings of support cradles are installed in line with the entry and exit idlers to avoid the creation of entrapment points. It is important that the bar directly under the steel chute or skirtboard wall be precisely aligned with the wing idlers at that point. The bars or rollers below the center of the trough are installed 0.5 to 1 inch (12 to 25 mm) below the unloaded belt's travel line to minimize friction when the belt is moving without cargo.

Cradles and Power Requirements

Belt support systems have a significant effect on the power requirements of a conveyor. Changes in belt support will have a particularly noticeable effect on short or under-powered systems. It is advisable to calculate the theoretical power requirements of proposed changes in belt support systems to make sure there is adequate conveyor drive power available to compensate for the additional friction placed on the conveyor.

The added horsepower is calculated by determining the added belt tension, using the standard methods recommended by CEMA. The coefficient of friction of the new (or projected) systems, multiplied by the load placed on the belt support system from belt weight, material load, and sealing system, equals the tension. (Estimates of the weight of the belt can be found in **Figure 6.53**). There is no need to allow for the removal of idlers, the incline of the conveyor, or other possible factors, as estimates provided by this method will in most cases be higher than the power consumption experienced in actual use. In applications where there is consistently a lubricant such as water present, the actual power requirements may be one-half or even less than the amount estimated through these calculations.

In *Belt Conveyors for Bulk Materials*, CEMA details a relatively complex formula for determining conveyor belt tension and power requirements. Current computer software offers similar equations (and will perform the calculation for you).

CEMA Standard 575-2000 introduces a new variable to the power equation published in *Belt Conveyors for Bulk Materials*. This factor is $T_e(IB)$; its formula is published as:

$$T_e(IB) = L_b [(W_m + W_b) \times f]$$

L_b = Length of Impact Bed, in feet
W_m = Weight of Bulk Material, per foot
W_b = Weight of Belt, per foot
f = Coefficient of friction

The net change in tension is $T_e(IB)$ minus the tension–$T_e(TI)$–from the idlers replaced by the impact bed.

Estimating Power Requirements

To estimate the additional power required by the installation of new belt support systems, you would need to know or estimate the following:

		Imperial	**Metric**
V	**Belt Speed**	Feet Per Minute (fpm)	Meters Per Second (m/sec)
Q	**Weight of Material Conveyed**	Tons Per Hour (tph)	Metric Tons Per Hour (mtph)
W_b	**Weight of Belt**	Pounds Per Foot (lb/ft)	Kilograms Per Meter (kg/m)

(If unknown, see Figure 6.53. Approximate Weight of Belt)

W_m	**Weight of Material on Belt**	Pounds Per Foot (lb/ft)	Kilograms Per Meter (kg/m)
	To determine, use this formula:	$\dfrac{33.33 \times Q}{V}$	$\dfrac{0.278 \times Q}{V}$

If only "side rail" support is to be installed, use Sealing Support Factor of 1/10 (0.1) of the weight of loaded belt ($W_b + W_m$), as this belt support system will contact only a portion of the belt surface, and the weight of the load is below the sealing supports.

L_b	**Length of Belt Support** (per side of the conveyor)	Feet (ft)	Meters (m)
T_{rs}	**Rubber Sealing Strip Load**	3 lb/ft per side	4.46 kg/m per side

(as identified by CEMA in *Belt Conveyors for Bulk Materials*, 5th Edition, page 102).
This number will vary greatly depending on component design and maintenance practices.

f **Friction**
Dry coefficient of friction of new belt support system against a moving dry conveyor belt. CEMA Standard 575 lists these estimates for **f**.

Material	f Static	f Dynamic
UHMW PE	0.5	0.33
Polyurethane	1.0	0.66

Precise information for the materials used in a given component should be available from the manufacturer.

Estimating Additional Power Requirements

Procedure

Complete both Step 1 and Step 2 below using appropriate formulas:

IMPERIAL

Step 1: *To determine Added Load on the Drive, use the following formulas:*

• For Sealing Support (side rail) only:

Weight of Belt and Load x Side Seal Factor + Skirting Load = Added Load

$$(W_b \times L_b \times 0.1) + (T_{rs} \times 2L_b) = \text{Added Load (lbs)}$$

• For Full Belt (impact cradle) support:

Weight of Belt + Skirting Load + Material Load On Belt = Added Load

$$(W_b \times L_b) + (T_{rs} \times 2L_b) + \frac{Q \times L_b \times 33.33}{V} = \text{Added Load (lbs)}$$

The constant 33.33 in the Imperial formula converts tons/hour to lbs.

Step 2: *To determine additional Horsepower Requirements, use the following formula:*

$$\text{Horsepower} = \frac{\text{Added Load (from Step 1)} \times V \times f}{33,000}$$

The constant 33,000 converts ft-lb/min to horsepower.

METRIC

Step 1: *To determine Added Load on the Drive, use the following formulas:*

• For Sealing Support (side rail) only

Weight of Belt and Load x Side Seal Factor + Skirting Load = Added Load

$$(W_b \times L_b \times 0.1) + (T_{rs} \times 2L_b) = \text{Added Load (kg)}$$

• For Full Belt (impact cradle) support:

Weight of Belt + Skirting Load + Material Load On Belt = Added Load

$$(W_b \times L_b) + (T_{rs} \times 2L_b) + \frac{Q \times L_b \times 0.278}{V} = \text{Added Load (kg)}$$

The constant 0.278 in the Metric formula converts metric tons to kg.

Step 2: *To determine additional Horsepower Requirements, use the following formula:*

Watts = Added Load (from Step 1) x V x f x 9.81

The constant 9.81 converts kg-m/min to watts.

To estimate the total power required for the conveyor, add the results of the above calculation to the current power requirements for the conveyor system.

Data for Sample Calculations
(Problem A & Problem B)

Here are the specifications for a typical conveyor to be examined in the following sample problems:

		Imperial Units	Metric Units
Material Conveyed		Wood Chips	
Material Density		19-30 lb/ft^3	304-480 kg/m^3
Belt Width		36 inches	914 mm
Belt Speed	V	250 FPM	1.27 m/sec
Tonnage Conveyed	Q	300 tph	272 mtph
Average Weight of Belt	W_b	9 lb/ft	13.4 kg/m
(per **Figure 6.53**.)			
Weight of Material on Belt, er foot	W_m	40 lb/ft	59.5 kg/m
Length of the Area to Be Supported	L_b	16 feet	4.88 meters
Skirting Seal Load	T_{rs}	3 lb/ft	4.46 kg/m
Friction Factor	f		
of Sealing Support System			0.5
of Impact Cradle Support System			0.5

Belt Support 6

Problem A

How much additional power would be required if a "side rail" sealing support system were installed for the full length of the transfer point on the conveyor identified? (This installation is seen in **Figure 6.54**.)
For this calculation we will ignore the reduction in idler friction.

Imperial Units

STEP 1:   Weight of Belting = (W_b x L_b x 0.1)   9 x 16 X 0.1   =   14.4 lb
plus   Skirting Load = (T_{rs} x $2L_b$)   3 x (2 x 16)   =   96.0 lb
                                                         Total Added Load: = 110.4 lb

STEP 2: $\dfrac{\text{Added Load \textit{(from Step 1)} x V x f}}{33{,}000}$ $\dfrac{110.4 \times 250 \times .05}{33{,}000}$ =   0.418 hp

                                                         Additional Power Required = 0.42 hp

Metric Units

STEP 1:   Weight of Belting = (W_b x L_b x 0.1)   13.4 x 4.88 x 0.1   =   6.54 kg
plus   SkirtingLoad = (T_{rs} x $2L_b$)   4.46 x (2 x 4.88)   =   43.53 kg
                                                         Total Added Load   =   50.10 kg

STEP 2:

Watts = Added Load (from Step 1) x V x f X 9.81
50.1 x 1.2 x 0.5 x 9.81   = 312.09 watts
                                                         Additional Power Required = ~ 0.31 kW

Solution:

To install a "side rail" sealing support system the full length of the transfer point on the conveyor as specified **(Figure 6.54)** would require approximately 0.42 horsepower or 312 watts (~0.3 kilowatts) of additional power.

Note: Remember to allow for the reduction in idler load, by subtracting the Te (TI) from the idlers removed. This will be particularly important on longer "side rail" support installations.

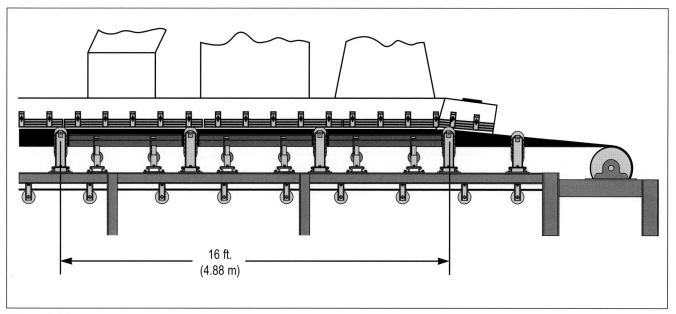

Figure 6.54

Conveyor as described in Problem A, featuring "side rail" sealing support only.

Problem B

To install a full belt width (impact) cradle on the entire length of the transfer point on the specified conveyor (as seen in **Figure 6.55.**) would require how much additional power?
For this calculation we will ignore the reduction in idler friction.

Imperial Units

STEP 1:

Weight of Belting ($W_b \times L_b$)		9 x 16	=	144.0 lb
plus Skirting Load = $T_{rs} \times 2L_b$		3 x (2 x 16)	=	96.0 lb
plus Material Load on Belt = $\dfrac{Q \times L_b \times 33.33}{V}$		$\dfrac{300 \times 16 \times 33.33}{250}$	=	879.9 lb
		Total Added Load	=	879.9 lb

STEP 2: $\dfrac{\text{Added Load (from Step 1)} \times V \times f}{33,000}$ $\dfrac{879.9 \times 250 \times 0.5}{33,000}$ = 3.33 hp

Metric Units

STEP 1:

Weight of Belting = ($W_b \times L_b$)		(13.4 x 4.88)	=	65.4 kg
plus Skirting Load = $T_{rs} \times 2L_b$		4.46 x (2 x 4.88)	=	43.5 kg
plus Material Load on Belt = $\dfrac{Q \times L_b \times 0.278}{V}$		$\dfrac{272 \times 4.88 \times 0.278}{1.27}$	=	290.6 kg
		Added Load	=	399.5 kg

STEP 2:

Watts = Added Load (from Step 1) x V x f X 9.81   399.5 x 1.27 . 0.5 x 9.81 = 2488.6 Watts (2760.2249)
= 2.5 kW

Solution:
To install a full belt width (impact) cradle on the entire length of the transfer point on the above conveyor **(Figure 6.55)** would add an additional 3.33 horsepower or 2488.6 watts (~2.5 kilowatts) to the conveyor's operating power requirement.

Note:
Remember to allow for the reduction in idler load, by subtracting the Te (Tl) for all idlers removed. This will be particularly important with installation of multiple impact cradles.

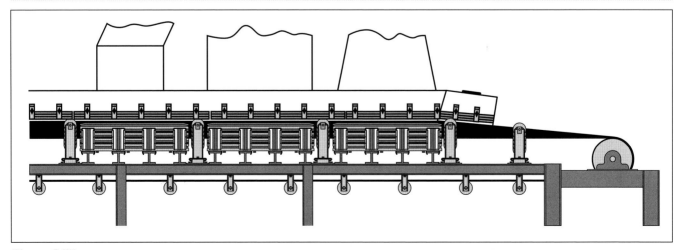

Figure 6.55

Conveyor as described in Problem B, featuring full belt width (impact cradle) support.

Pay Now, Or Pay (More) Later

The installation of improved belt support systems will increase the conveyor drive power requirements. However, the true implications of this system are seen when it is compared to the power consumption of a conveyor where idler bearings drag or the idlers themselves build up with material due to transfer point spillage induced by belt sag.

As demonstrated by R. Todd Swinderman in the paper "The Conveyor Drive Power Consumption of Belt Cleaners," fugitive material can also impair the operation of conveyor systems, increasing power consumption significantly. For example, Swinderman calculates that a single frozen impact idler set would require approximately 1.6 additional horsepower (1.2 kW), while a seized steel idler set can demand as much as 0.36 additional horsepower (0.27 kW). One idler with a one inch (25 mm) accumulation of material would add 0.43 additional horsepower (0.32 kW) to the conveyor's drive requirements.

The point is it makes more sense to design this power consumption into the system to prevent the spillage, rather than allow the spillage to happen and try to compensate later through increases in motor size and power consumption.

The use of improved belt support and sealing techniques will place an additional requirement on conveyor drive systems. However, these additional requirements (and costs) will seem minor when compared to the power consumed by operating with one "frozen" idler or several idlers operating with a material accumulation.

By implementing the proper belt support systems, a plant can prevent the many and more costly problems that arise from the escape of fugitive material. The costs for installation and operation of proper belt support systems represent an investment in efficiency.

chapter 7

Skirtboard

"The skirtboard of each transfer point must be engineered to match the characteristics of the material and the conveyor..."

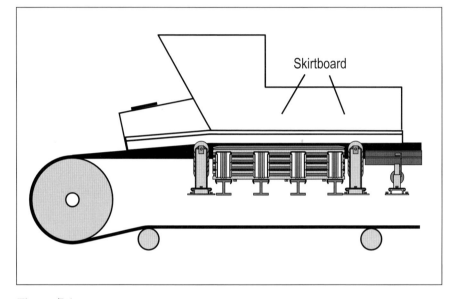

Figure 7.1
Skirtboard is the horizontal extension of the loading chute.

If the chute contains the material as it moves in a downward dimension, the skirtboard contains the material in a horizontal dimension after the material is loaded onto the belt. **(Figure 7.1)** Also referred to as skirt plates, steel skirting, or sometimes merely as chute or chutework, a skirtboard is a strip of wood or metal extending out from the load point on either side of the belt. **(Figure 7.2)**

The purpose of skirtboard is to keep the load on the conveyor, preventing the material from spilling over the belt edge until the load is settled and has reached belt speed. The skirtboard of each transfer point must be engineered to match the characteristics of the material and the conveyor and the way the transfer point is loaded and used.

In this volume the word "skirtboard" is used for these metal or wood extensions of the chute. The terms "rubber skirting," "skirtboard seal," "side wipers," "dust seal," and "edge seal" refer to the rubber or plastic strip installed below the metal skirtboard to prevent the escape of fines.

The Role of Skirtboard in Control of Fugitive Material

The skirtboard and the wear liner placed inside the skirtboard combine with the elastomer sealing system to form a multiple-layer seal. The elastomer seal strips cannot and should not be expected to withstand

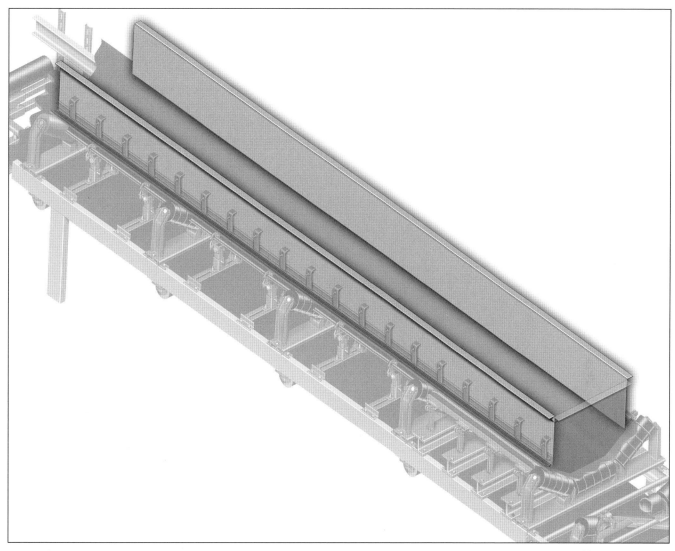

Figure 7.2
Skirtboard confines and shapes the cargo until it has settled into the desired profile.

significant material side pressures. The sealing system should only be expected to keep dust and fines from escaping. The steel of the skirtboard and wear liner form the first lines of defense to contain this material and prevent the material head pressure from prematurely wearing the sealing system. *Sealing systems are discussed in Section 9.*

In addition, the skirtboard serves as the foundation and basic enclosure for an effective dust management system.

Skirtboard Length

Theoretically, the skirtboards should extend in the belt's direction of travel past the point where the material load has fully settled into the profile it will maintain for the remainder of its journey on the conveyor.

Sometimes the load never becomes completely stable, and skirtboard is required for the entire length of the conveyor. This is most common with very fine materials that are easily made airborne or materials that tend to roll. Belt feeders, which are often short in length and loaded to the full width of the belt, are also commonly skirted for their full length.

In *Belt Conveyors for Bulk Materials*, the Conveyor Equipment Manufacturers Association (CEMA) notes that the length required for the skirtboards–when loaded in the direction of travel–is "a function of the difference between the velocity of the material at the moment it reaches the belt and the belt speed."

It must be understood that this distance refers to the length of skirtboard beyond the impact zone. The impact zone is equal in length to the loading chute in the direction of material flow.

The engineering decision as to the length of the skirtboard is often based on rules of thumb. In some cases these rules underestimate the necessary length. CEMA recommends two feet of skirtboard for every 100

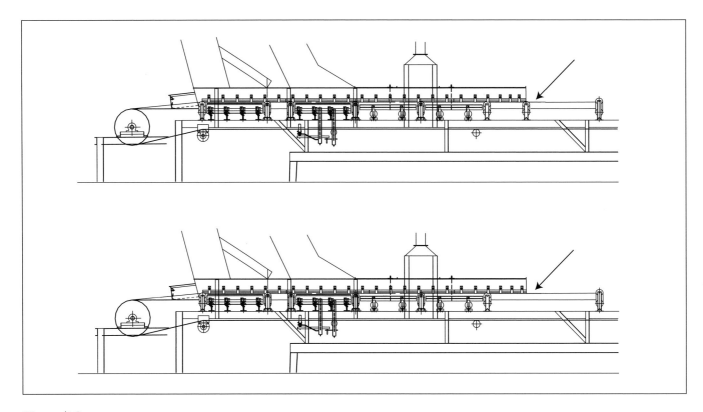

Figure 7.3
It is preferable for skirtboard to end above an idler (top), rather than between idlers (bottom).

feet-per-minute (.6 meters for every meter/second) of belt speed, with a minimum length requirement of not less than three feet (.9 meter). While the CEMA recommendation may be adequate for some materials that are easily conveyed, it is insufficient for materials that require dust collection or suppression, for materials that tend to segregate or roll, or for transfer points that are not in line with the direction of belt travel.

One recent specification, apparently based on the CEMA recommendation, requires:

> The length of the skirted area after the impact zone shall be two feet for every 100 feet-per-minute (600 mm for every 0.5 m/sec) of belt speed. This applies given the following conditions:
> - Maximum belt speed is 700 feet per minute (3.5 m/sec).
> - Maximum angle of conveyor incline is seven degrees.
> - Maximum air flow is 1000 cfm (28,300 l/min).
>
> If operating parameters exceed any of the above, the length of the skirtboard shall be three feet per 100-feet-per-minute (900 mm for every 0.5 m/sec) of belt speed.

Other authorities offer their own formulas. In *Mechanical Conveyors for Bulk Solids*, Hendrik Colijn notes, "Skirt plates should extend 8 to 12 feet beyond the point where the main material stream flows onto the belt." The U.S. Army Corps of Engineers technical manual, "Design of Steam Boiler Plants," provides a different formula: "Skirtboard length will be five feet from impact point, plus one foot for every 100 fpm of belt speed."

Faced with these different formulas, an engineer could do worse than to perform each of the calculations and pick the median value. In the interest of erring on the side of having more skirtboard than the minimum requirement–and so having better control of spillage and dust–it would be better to specify the longest length yielded by the above formulas.

To minimize the possibility that fluctuations in the belt travel line will push the belt up into a steel edge, skirtboard should end above an idler, rather than between idlers. This in itself may provide some guidance in determining the overall length of the skirtboard. **(Figure 7.3)**

What may provide a more telling answer for skirtboard length is the need for enclosing the dust suppression and collection systems as discussed in this volume. The walls of a dust control enclosure can effectively serve as skirtboard, with the length necessary for effective dust control systems generally providing more than what is required for load stabilization.

The only penalties for increasing the length of the skirtboard will be the additional maintenance cost for the longer liners and seals, a minimal increase in the cost of the steel for the walls, and an increase in the conveyor's horsepower requirements. The extra power consumption results from the added friction created by the additional length of sealing strip required to seal the extended skirtboard. This is usually a modest increase. *See Chapter 9 for information on the power consumption of edge sealing strips.*

Skirtboard Width

The width of the skirtboard–the distance between the two sides of the skirtboard–is often determined by the space needed to establish an effective seal on the outside of the skirtboard. A distance of four to six inches (100 to 150 mm) outside the chute per side is

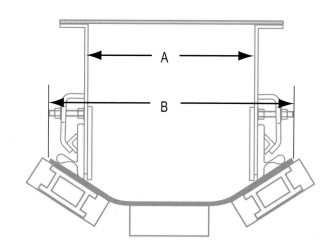

A - Maximum recommended chutewall width (based on trough angle)
B - Effective belt width (in troughed position)

Recommended Load Zone Design

Belt Width in. (mm)	Trough Angle	Effective Belt Width in. (mm)	Recommended Chute Width in. (mm)
18 (400-500)	0°	18.0 (457)	6.0 (152)
	20°	17.3 (439)	5.5 (140)
	35°	16.2 (411)	3.7 (94)
	45°	14.9 (378)	2.5 (64)
24 (500-650)	0°	24.0 (610)	12.0 (305)
	20°	23.0 (584)	11.8 (300)
	35°	21.6 (549)	11.8 (300)
	45°	19.9 (505)	11.4 (290)
30 (650-800)	0°	30.0 (762)	18.0 (457)
	20°	28.8 (732)	17.6 (447)
	35°	27.0 (686)	17.2 (437)
	45°	24.9 (632)	16.4 (417)
36 (800-1000)	0°	36.0 (915)	24.0 (610)
	20°	34.6 (879)	23.4 (594)
	35°	32.4 (823)	22.6 (574)
	45°	29.9 (759)	21.4 (544)
42 (1000-1200)	0°	42.0 (1067)	30.0 (762)
	20°	40.3 (1024)	29.1 (739)
	35°	37.8 (960)	28.0 (711)
	45°	34.9 (886)	26.4 (671)
48 (1200-1400)	0°	48.0 (1219)	36.0 (914)
	20°	46.1 (1171)	34.9 (886)
	35°	43.2 (1097)	33.4 (848)
	45°	39.8 (1011)	31.3 (795)
54 (1400-1600)	0°	54.0 (1372)	42.0 (1067)
	20°	51.8 (1316)	40.6 (1031)
	35°	48.6 (1234)	38.8 (986)
	45°	44.8 (1138)	36.3 (922)
60 (1600-1800)	0°	60.0 (1524)	48.0 (1219)
	20°	57.6 (1463)	46.4 (1179)
	35°	53.4 (1356)	43.6 (1107)
	45°	49.2 (1250)	40.7 (1034)
72 (1800-2000)	0°	72.0 (1829)	60.0 (1524)
	20°	69.1 (1755)	57.9 (1471)
	35°	64.1 (1628)	54.3 (1379)
	45°	59.0 (1499)	50.5 (1283)
84 (2000-2200)	0°	84.0 (2134)	72.0 (1829)
	20°	80.6 (2032)	69.4 (1763)
	35°	74.8 (1900)	65.0 (1650)
	45°	68.9 (1750)	60.4 (1534)
96 (2200-2400)	0°	96.0 (2438)	84.0 (2134)
	20°	92.2 (2342)	81.0 (2057)
	35°	85.4 (2169)	75.6 (1919)
	45°	78.7 (1999)	70.2 (1783)

(Dimensions were determined by calculation rather than field measurements. English measurements are rounded to one decimal; metric measurements are a conversion of English units. Thickness of steel in chute or skirtboard is not considered.)

Figure 7.4

Effective Belt Width (Available Load-Carrying Space).

generally sufficient for effective sealing on the inclined edges of a troughed belt.

In the reference *Belt Conveyors for Bulk Materials*, CEMA specifies the standard edge distance (the distance from the edge of the material load to the outside edge of the belt) per side as:

> 0.055b + 0.9 inch = Standard Edge Distance,
> where b is width of the belt in inches.

The metric conversion of this formula is

> 0.055b + 23 mm = Standard Edge Distance,
> where b is width of belt in mm.

This formula provides enough space to seal larger conveyors–those over 48 inches (1200 mm) wide–but is less useful on narrower belts. On a 96-inch (2400-mm) belt, the formula leaves 6.18 inches (157 mm) on each side; this is an acceptable distance, providing both effective sealing room and sufficient load capacity. However, using this formula on a 48-inch (1200-mm) belt leaves 3.54 inches (90 mm) on each side. This provides an edge distance that is barely sealable. On a 24-inch (600-mm) belt, this formula leaves only 2.22 inches (56 mm) per side, which is insufficient to allow for sealing or to accommodate minor mistracking.

In its discussion of "Conveyor Loading and Discharge," the CEMA book offers another formula: "The maximum distance between skirtboards is customarily two-thirds the width of a troughed belt." This CEMA formula can be restated as "one-sixth of the belt width should be provided on the outside of the skirtboard area on each side of the belt to allow for proper sealing."

This CEMA "Two-Thirds of Belt Width" formula provides an easy-to-remember, easy-to-calculate determination of skirtboard width. But the variables of real life intrude on this simplicity and limit its useful application.

The variations occur with the changes in a belt's "effective width"–the space actually available for carrying material–created by forming the belt into a trough. As the trough angle increases, the effective belt width decreases. Consequently, if the chute width is kept at two-thirds of the actual (flat) belt width, and this skirtboard width is subtracted from the effective belt width, there may not be sufficient edge distance outside the skirtboard for effective sealing.

Figure 7.4 specifies the effective belt width of various sizes of belts at typical troughing angles.

In the mid-range of belt sizes, from 42 to 60 inches (1050 to 1500 mm), the CEMA "Two-Thirds of Belt Width" formula provides an acceptable space for sealing.

But for belt sizes above 60 inches (1500 mm)–especially with the lower troughing angles–the CEMA formula consumes too much load-carrying capacity by leaving too much belt outside the load carrying area. For example, on a 72-inch (1800-mm) belt with a 20° trough, the "Two-Thirds" formula provides a chute width of 48 inches (1200). But because the effective belt width is 69.1 inches (1755 mm), there is a distance of over 21 inches (550 mm) left outside the load carrying area. That's more than 10 inches (250 mm) of belt per side. This is probably more space than is required for sealing, and thus wastes the belt's carrying capacity.

For the sizes of 36 inches (900 mm) and below, the "Two-Thirds" formula does not leave sufficient space outside the skirtboard to allow effective sealing of belt edges. For a 36-inch belt, the skirtboard width using the CEMA "Two-Thirds of Belt Width" formula is 24 inches (600 mm). But if this 36-inch belt is troughed at 45°, a skirtboard of this width leaves only 5.9 inches (150 mm) of belting outside the chute. There is less than three inches (76 mm) per side, which is not enough room to establish an effective seal.

To determine the maximum skirtboard width using **Figure 7.4**, subtract a minimum of eight inches (200 mm)–or four inches (100 mm) per side–from the Effective Belt Width for belts 36 inches and above. For belts from 18 to 30 inches (450 to 750 mm) subtract a minimum of six inches (150mm) or three inches (75mm) per side. This will allow for installation of an effective sealing system.

Both the "standard edge distance" formula and the "Two-Thirds of Belt Width" formula should be viewed as providing rough approximations, rather than finished calculations. In these cases, it is best to crosscheck the formula through several procedures to make sure the optimum solution is determined.

Skirtboard Height

The height of the skirtboard can be based on the volume of the material load and the dimensions of the largest lumps and on the volume and speed of air produced by the transfer point.

The skirtboard should be tall enough to contain the material load when the belt is operating at normal capacity. As the size of lumps included in the load goes up, so must skirtboard height; at minimum, the height must be sufficient to contain the largest pieces. It is better if the skirtboard probably is tall enough to contain two of the largest pieces piled on top of each other.

In *Belt Conveyors for Bulk Materials*, CEMA has published a table of minimum uncovered skirtboard heights. In summary, it specifies approximately 12 inches (300 mm) is tall enough for particles of two inches (50 mm) or smaller, carried on flat or 20° troughed belts up to 72 inches (1800 mm) wide, or for 35° and 45° troughed belts up to 48 inches (1200 mm) wide. The table specifies skirtboards up to 32.5 inches (825 mm) in height for belts as wide as 96 inches (2400 mm) with lumps up to 18 inches (450 mm).

For materials that may create a dust problem, it is good practice to increase the height of the skirted area to provide a plenum. This area will contain and still the dust-laden air so that the particles can settle back onto the cargo of the conveyor.

For maximum dust control, the chute (skirt) walls must be high enough to furnish a cross-sectional area in the load zone that provides a maximum air velocity above the product bed of less than 250 feet per minute (1.25 meter per second).

This larger area provides an ample plenum chamber to accommodate the positive pressures of air movement without material blowing out of the enclosure. In many circumstances, skirtboard height must be increased to 24 inches (600 mm) or more. This gives air currents plenty of room to slow, which will minimize the amount of dust entrained. Chute walls need to be high enough so that dust exhaust pick-ups mounted in the covering (or roof) of the skirtboard do not pull fines off the pile. If the walls are too low, the dust collectors will vacuum fine material from the piles and may plug themselves with material. If the walls are not high enough, energy will be wasted removing dust that would have shortly settled on its own, and the dust collection system will be larger (and more expensive) than necessary.

Factors including belt speed, material characteristics, drop height, and air speed at the discharge should be considered when evaluating skirtboard height.

"Double-height" skirtboards of 24 inches (600 mm) are frequently recommended for materials prone to high levels of dust. Skirtboard that is 30 to 36 inches (750 to 900 mm) high should be used if the belt speed is above 750 fpm (3.8 m/s), and the material is prone to produce dust.

Skirtboard Coverings

Covering the skirtboard with a steel or fabric system is recommended for a dust control. The covering of the skirtboard is required to create the plenum needed to allow dust to settle and air movements to be stilled. As noted in the sections on control of air movement, dust collection, and dust suppression, a large plenum is useful in controlling the clouds of dust driven off by the forces of transferring the body of material.

In addition, placing a "roof" over the skirtboard will contain the occasional lump of material that through some freak circumstance comes through the loading chute onto the belt with force sufficient to "bounce" it completely off the belt.

Unless a specific reason exists *not* to cover it, the skirtboard should be enclosed. (Sufficient openings should be provided for service and inspection; these openings should be provided with doors to prevent material escape and minimize the outflow of air.)

CEMA recommends that these covers should be slanted down from the chute to the skirtboard, allowing for material that is not yet moving at belt speed to avoid material jams. A preferred approach would be to make the skirtboard tall enough so that there is no possibility of this blockage. By keeping the skirtboard tall, you accomplish two things: avoid material jams and provide a large area to dissipate air velocity and let dust settle.

In most cases, steel is best. Fabric is often used to connect vibrating equipment to stationary skirtboards.

The covering should be designed so that it will support the weight of a worker, or it should be guarded and marked with "no step" warnings so nobody falls through the covering.

Skirtboard-To-Belt Distance

Even under the most ideal conditions, steel skirtboards can be hazardous to the belt. Fluctuations in its line of travel allow the belt to move up against the steel to be gouged or cut. In addition, material can wedge under the skirtboard to abrade the belt's surface.

It is critical to raise the skirtboards far enough above the conveyor so they never come in contact with the belt cover. However, skirtboard is often installed with a clearance of several inches above the belt to facilitate belt replacement. When the steel is placed this far above the belt surface, it is virtually impossible to provide an effective seal.

Once an ineffective seal is established, it perpetuates itself. Material leaks out, accumulating on idlers, leading to mistracking and other problems that result in an

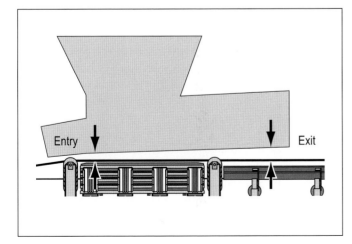

Figure 7.5
The distance between the belt and the bottom of the skirtboard should open gradually toward the exit of the transfer point.

unstable belt line. The belt flexes up and down and wanders from side to side. Plant engineers and maintenance staffs, mindful of the need to prevent the belt from coming into contact with the chutework, increase the belt-to-skirting clearance. This increases the difficulty in sealing and results in increased spillage, continuing the vicious circle.

The gap between the belt and the skirtboard should be as small as possible. The closer the steel and belt are together, the easier it is to maintain a seal between them. As an example, the lower edges of the skirt plates should be positioned 1/4 inch (6 mm) above the belt at the belt's entry into the loading zone. This dimension should be uniformly increased in the direction of belt travel to 3/8 inch (9 mm). **(Figure 7.5)** This close clearance cannot be accomplished unless the belt travel is stabilized within a plus-or-minus tolerance of 1/16 inch (1.5 mm) at the entry (tail pulley) end of the chute.

It is critical to provide relief in the direction of belt travel. The gap under the steel should form a wedge-shaped opening to allow conveyed material to ride along the steel skirting and sealing rubber, rather than become wedged into an opening by the ceaseless force of belt motion. The skirtboard should open gradually, both horizontally and vertically, from the loading point in the direction of belt travel to permit entrapped material to free itself.

With the steel positioned close to the belt line, it is critical to the safety of the belt that the belt be prevented from rising up off the idlers during conveyor start up. This is one reason why the elevation of the tail pulley (generally done in the interest of shortening the transition area) is not a good idea, as this practice encourages the belt to rise. It is important that the belting specifications and tension be calculated correctly to minimize the risk of the belting lifting off the idlers. Hold-down rollers can be installed to keep the belt on the idlers.

Rough bottom edges or warped steel can create difficult conditions, capturing material to increase the drag on the conveyor drive and/or abrade the belt surface. Likewise, ceramic blocks or wear plates must be carefully installed to avoid jagged or saw-toothed edges that can trap material or damage the belt. The rule is to maintain a smooth flow surface on the bottom edge of the skirtboard and eliminate all entrapment points. Again, the skirtboard should gradually taper open as the belt moves toward the exit of the loading zone to allow relief of pressure and entrapped material. Skirtboard steel and chute liners must be installed very carefully, with all seams well matched.

The gap left between the skirt and the belt surface should be sealed by a flexible, replaceable elastomer sealing system applied to the outside of the skirtboard.

Figure 7.6
Skirtboard supports should be far enough above the belt to allow access to the sealing system.

Conveyor Characteristics	A-Frame Angle Iron Size
Below 750 fpm (3.7 m/sec) or 54 inches (1400 mm) wide.	2" x 2" x 3/16" (50 x 50 x 5 mm)
Up to 750 fpm (3.7 m/sec) or 54 inches (1400 mm) wide	3" x 3" x 1/4" (75 x 75 x 6.4 mm)
Up to 1000 fpm (5 m/sec) or 72 inches (1800 mm) wide	3" x 3" x 3/8" (75 x 75 x 9.5 mm)

Figure 7.7
Recommended Angle Iron Sizes for Skirtboard Supports.

Sealing systems are discussed in detail in Chapter 9: Edge Sealing.

Skirtboard Construction

The strength and stability of the skirtboard are very important to its success. Many times conveyor skirtboard is supported by cantilever brackets that are not rigid enough to withstand the impact of material or the vibration of equipment. The thickness of the skirtboard must be sufficient to withstand side pressures that may occur when the chute becomes plugged or the belt rolls backward. As it is located close to the belt, any movement of the skirtboard must be prevented to minimize the risks of damage.

Except in very light applications, the minimum thickness of mild steel used for skirtboard construction should be 1/4 inch (6 mm). On belts moving at over 750 fpm (3.7 m/sec) or 54 inches (1300 mm) or more wide, the minimum thickness should be 3/8 inch (9 mm). For applications over 1000 fpm (5 m/sec) or 72 inches (1800 mm) wide, the minimum thickness should be 1/2 inch (12 mm).

Skirtboard should be supported at regular intervals with structural steel. The most common support design is an angle iron "A-frame" installed on approximately the same centers as the carrying idlers. **(Figure 7.6)** These "A-frames" should be rigid and well-gusseted, and installed far enough above the belt to allow easy access for the adjustment or replacement of the skirboard seal.

The minimum sizes of angle iron used to construct these "A-frames" are shown in **Figure 7.7.** These specifications are best suited for low-density, free-flowing materials. For belt feeders, or handling high-density materials like ore handling or concentrate, heavier steel and closer spacing are required.

The "A-frames" should be installed on approximately 48-inch (1.2-meter) centers, without interfering with the spacing of belt support cradles and idlers. One A-frame should be positioned at the beginning of the chutework and another at the end. Closer spacing should be considered in the conveyor's impact zone to the extent of doubling the support structure.

It is important that proper clearance be provided between the bottom of the skirtboard supports and the belt in order to allow room for the installation and maintenance of a skirtboard seal and clamp system. Ideally the distance should be 12 inches (300 mm); a minimum would be 9 inches (230 mm). The greater the clearance, the greater the need for properly bracing the skirtboard to prevent its movement.

If there is dynamic vibration in the system caused from either belt movement or other operating machinery such as breakers, crushers, or screens, the skirtboard may need to be isolated from it. Structural adjustments may be required to prevent skirt steel from picking up these vibrations and oscillating at an amplitude that prevents effective sealing.

Skirtboard for Conveyors with Multiple Loading Points

Where a belt is loaded at more than one point along the conveyor's length, care must be used in the positioning of the skirtboard at the loading points. Because the load of the material tends to flatten and spread on the belt, skirtboard at the secondary load points must be designed to allow the previously loaded material to pass freely.

The use of a belt that is wider or more deeply troughed than normal will facilitate loading without spillage at additional loading points as the cargo will have been centered by the initial load zone.

Load centralizing idlers can be installed to consolidate the load and prepare it for entry into a second load zone with conventional skirtboard.

When loading points are relatively close together, it is often better to continue the skirtboards between the two loading points rather than use relatively short lengths only at the loading points. Continuous skirtboards are good insurance against spillage on belts with multiple load points.

Another approach would be to use an air-supported conveyor. These conveyors are uniquely suited for multiple load zones, as they require only a centered load, rather than conventional skirtboard or skirt seals.

chapter 8

Wear Liners

"This pressure, if uncontrolled, will push material fines and dust away from the center of the material pile, resulting in spillage."

Wear liner is a material installed on the interior of the skirtboard as a sacrificial surface. In plans for a low-spillage transfer point, it serves multiple purposes:

A. It provides an easily replaceable sacrificial wear surface to protect the walls of the chute and the skirtboard.
B. It helps center the material load inside the skirtboard and on the belt.
C. It prevents the material load from applying high side-forces onto the sealing strips, thereby improving the seal's service life.

In this text, the phrase "wear liner" is used to describe a lining put inside the transfer point's skirtboard; that is, the horizontal extension of the chute.

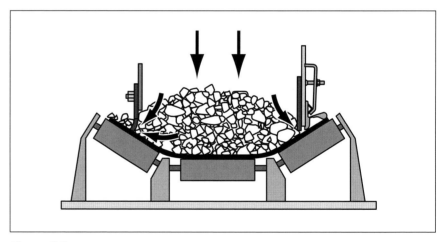

Figure 8.1

If not protected by wear liner, as shown on left, the elastomer sealing system is not strong enough to contain the forces that push material over the edge of the belt.

What Wear Liner Does

The forces of material dropping onto a belt inside a transfer area create tremendous outward pressures. **(Figure 8.1)** This pressure, if uncontrolled, will push material fines and dust away from the center of the material pile, resulting in spillage. For years, standard practice has been to apply a sealing strip of some kind at the bottom of the skirtboard to keep material from escaping. But if the material load exerts too much force against the sealing, the sealing system will be unable to control and withstand these powerful side pressures. The forces will push the sealing system away from the wall, up from the belt, or abrade it rapidly. The result is material spillage over the sides of the belt in the loading zone.

In addition, the rubber and other elastomers commonly used as sealing strips tend to retard the flow of the granular material as the material moves along the rubber surface. As this material on the outer edge of the load slows, the force of the material on top of the pile that is trying to achieve belt speed pushes more material toward the outside, magnifying the pressure against the sealing system.

Wear liners are installed inside the skirtboard to protect the skirting seal. They have the mission of divorcing the job of sealing from the load placement function. By creating a dam between the material pile and the edge-sealing strips, the wear liners greatly reduce the side-load forces from the material that reaches the sealing strips. With wear liners installed, the strips do not have to act as a wall to contain the material load but rather as a seal, a purpose for which they are much more suited. This arrangement improves the effectiveness and life expectancy of the sealing system while reducing the risk of damage by material entrapment.

There are only a few instances where the installation of a wear liner would not greatly enhance the sealability and life expectancy of a transfer point. These cases would be very lightly loaded belts or belts handling non-abrasive, low-density materials. In all other circumstances, properly installed and maintained wear liners will reduce the material side-loading forces and increase skirting life.

Wear Liner Configurations

Two styles of wear liner are commonly seen today: straight and deflector. **(Figure 8.2)**

Straight Wear Liner (Figure 8.3) has the capability of preventing side loading forces on the skirting seals without choking the chute and constricting material flow. It was originally used only on smaller size belts–36 inches (900 mm) or smaller–and belts that were running at capacity. Straight wear liner has also been used on all sizes of belts where impact is present, because it eliminates the risk of material bouncing and becoming wedged in the bend of deflector wear liner.

Current design thinking specifies straight wear liner under almost all circumstances. The real benefit of straight wear liner is that it provides improved life and improved sealing effectiveness without closing down the effective load area. In an era when more and more production is asked of fewer and fewer resources, it is important to maximize system capacity by utilizing the full width of the loading chute and conveyor belt. Straight wear liner is the best choice to meet both current and future material flow requirements.

Straight wear liner is also best for belts with multiple loading points, whether installed in one long transfer point or through several loading zones.

Deflector Wear Liner incorporates a bend so the bottom half of the liner is bent inward, toward the middle of the belt. **(Figure 8.4)** This angle provides a "free" area between the rubber skirting and the wear liner. This area is useful, as fines that have worked their way under the bottom edge of the wear liner still have an area on the belt to travel; they are not automatically ejected from the system. These particles are contained by the sealing strip and have a path to travel down the belt to the exit area of the transfer point. The fines that work their way out to challenge the rubber seal are relatively free of applied forces; they are isolated from the downward and outward force of the material load.

The drawback of deflector wear liner is that it reduces the effective cross-sectional area of the skirtboard area. This, in turn, reduces the volume of material that can

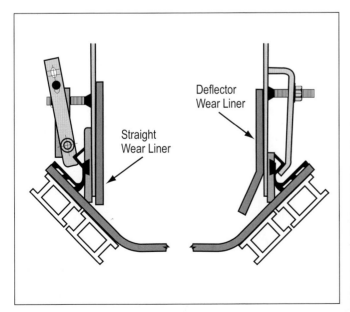

Figure 8.2

Left: Straight Wear Liner. Right: Defector Wear Liner.

Figure 8.3

Straight wear liner improves sealing without choking the effective load area.

pass through the transfer point, and consequently, may require adjustments in the dimensions to maintain flow capacity. This consideration is particularly important on smaller belts–less than 30 inches (750 mm) wide–or belts running near capacity. By closing down the load zone's cross section, deflector liner may also reduce the maximum allowable lump size, leading to material jams.

In addition, deflector wear liner should not be used in loading zones that face impact. In such applications, it faces higher wear and the opportunity for pieces of material to rebound off the belt and wedge themselves into the deflector liner's open bottom area, creating the risk of belt abrasion.

Spaced Wear Liner

A variation on wear liner installation technique is spaced wear liner. **(Figure 8.5)** This hybrid technique can be used in applications where mechanical dust collection is present. To assist in sealing, the liners are not installed directly onto the wall of the skirtboard but rather separated slightly from it. The space between the skirtboard and the wear liner inside it is used as negative pressure area. Fines and airborne dust in this area can be pulled from this space by the conveyor's dust collection system.

Obviously this technique is better suited for use on new conveyor systems, so the requirement for the "free area" can be engineered into the dimensions of the loading zone from the beginning. While the dimensions of this space are not large–one to two inches (25 to 50 mm) of free space on each side of the conveyor–it is much easier to allow for this space in the original design stage than to retrofit it into a transfer point. On most existing conveyors, the application of spaced wear liner will choke down usable space within the loading zone.

It is important in a spaced wear liner installation that the liner be installed so its top edge is well above the height of the material pile in the loading zone.

Wear Liner Selection Criteria

While the initial cost of a wear liner should be a significant consideration, it is more important that the material be selected on the basis of service life and performance. Factors that should also be considered include:
- Friction Coefficient
- Resistance to Material Adhesion
- Resistance to Abrasive Wear
- Resistance to Impact
- Resistance to Corrosion
- Attachment Method
- Installation Cost
- Maintenance Cost

Wear Liner Materials

Wear liners are typically supplied as sheets of material, often 72 inches (1829 mm) long, 8 inches (203 mm) high, and 0.5 inches (12 mm) thick. The liners can be furnished with pre-drilled holes to allow simplified field installation.

There are a number of materials suitable for use as wear liners. **(Figure 8.6)**

Mild Steel Wear Liner is commonly used on materials with very low abrasion or on belts with light loads or low operating hours. Materials such as sawdust, wood chips, foundry sand, and garbage would be good examples of material suitable for mild steel wear liners. In addition, projects with demands for low initial costs, but which require good short-term results, are also candidates for mild steel wear liner.

If the environment is damp or otherwise corrosive, the higher corrosion rate of mild steel may add additional friction to the material body in the loading zone.

Mild steel wear liner can be supplied in either the straight or deflector pattern.

Figure 8.4

Deflector wear liner creates a free area between the liner and the sealing system.

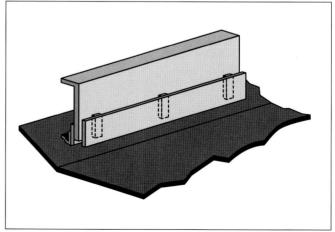

Figure 8.5

Spaced wear liner incorporates an open space behind the liner where dust collection can be applied.

COMPARISON OF LINER MATERIALS					
Lining Material	Initial Cost	Sliding Abrasion Resistance	Impact Resistance	Temperature Resistance	Low-Friction Quality
Mild Steel	Low	G	G	VG	-
Abrasion Resistant Plate	Medium	VG	G	VG	-
Stainless Steel	High	G	G	E	VG
Chromium Carbide Overlay	Medium	E	G	VG	VG
Rubber	High	G	E	-	-
Polyurethane	High	E	E	-	G
UHMW	Medium	G	-	-	E
Ceramic Tile					
Quarry Tiles	Low	G	-	G	G
Vitrified Tiles	Low	VG	-	VG	VG
Basalt Tiles	Medium	VG	G	VG	G
Alumina Tiles	High	E	G	E	G

Figure 8.6

Performance Comparison of Possible Wear Liner Materials. Ratings: E: Excellent; VG: Very Good; G: Good; –: Not Recommended.

Abrasion Resistant Plate Wear Liner (AR Plate) provides a much longer life than wear liner fabricated from mild steel. AR Plate is a good, all-around wear liner, capable of handling more abrasive materials such as sand, hard rock mining ores, and coal. The wear life may extend five to seven times longer than mild steel. AR Plate is available in either straight or deflector styles.

Ceramic-Faced Wear Liner is a good, long-term wear liner for continuously operating belts carrying highly abrasive material. A mild steel backing plate faced with ceramic blocks is a good choice in these circumstances. These ceramic blocks are glued and/or plug welded to the mild steel backing, usually on the bottom four inches (100 mm) of the plate. On more heavily-loaded belts, the ceramic blocks can also be applied higher up the backing plate to reduce wear.

Ceramic-faced wear liner has been shown to work well with coal and wood chips. Due to the brittle nature of ceramic tiles, this type of wear liner should not be used in areas of severe impact. Ceramic-faced wear liner can be supplied in both straight and deflector styles.

Any time liners are faced with castable materials, whether ceramic or alloys such as magnesium steel, extreme care must be taken to align the blocks during their installation on the steel plate. The bottom edge of the installation must be positioned with care to avoid pinch points and "stair steps" that can trap material.

Stainless Steel Wear Liner is a low-friction choice that falls between mild steel and AR Plate in abrasion resistance. The chemical resistance of stainless steel is often required for applications where the possibility of corrosion on mild steel or AR plate exists. Stainless steel wear liner can be supplied in both straight and deflector styles.

Chromium Carbide Overlay is a very hard material suitable for conveyors seeing very high levels of abrasion. Alone, chromium carbide is very brittle, so it is overlaid onto a backing plate for installation. The backing plate can be of mild or stainless steel, depending on application requirements. The hard facing is between 53 and 65 Rockwell "C", and some overlay materials "work harden" (be abused by the material so they become even harder) to 75 Rockwell "C". Also referred to as "clad plate" these materials are available in two designations: single weld or double weld pass. For wear liner applications, the double weld pass grade is typically used. This material is not suited for high impact, and as a result, is used only in the straight style.

Plastic Wear Liners are a more recent development. Recently, wear liners composed of ultra-high-molecular-weight polyethylene (UHMW) or urethane have been installed. In many of these installations, the liner sits directly on the belt to control extremely fine, dusty materials. Slotted holes in the liner panels allow adjustment to keep the wear liner in contact with the belt.

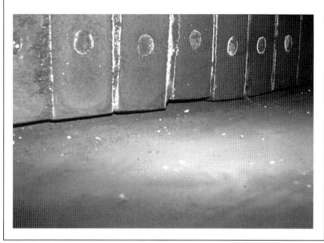

Figure 8.7

(Top) Poorly aligned wear liner creates material entrapment points.
(Bottom) The saw-tooth edge of this wear liner will capture material that can gouge the belt.

Figure 8.8

Wear liners should be installed in a straight line that gradually opens toward the exit of the transfer point.

Applications of UHMW as a wear liner show success with fine, powdery products such as sand, fly ash, and electric arc furnace (EAF) dust. In addition, as UHMW is accepted by the U.S. Food and Drug Administration, it is suitable for use with powdery foodstuffs. Urethane liners are now being used successfully in gold mining and ore processing applications for their light weight and ease of replacement.

Plastic materials have been applied only as straight wear liners; the abrasion that would be seen in applications as a deflector design would dramatically reduce the service life. Care should be taken not to install plastic liners in conditions that exceed the material's service temperature, or see a high belt speed that can raise the liner's temperature to a softening point that would shorten the material life.

Many of these materials are also suitable for lining curved chutes for applications where there is a need for wear resistance or reduced friction. Examples would include ceramic tiles or AR plate used as a liner in a curved chute handling coal or UHMW used in a chute for wood chips.

Application of Wear Liner

With the exception of UHMW wear liner, all wear liner systems must be installed with a relieving angle from the entry area opening toward the exit area of the transfer point. The distance above the belt will vary with the product size. As with the skirtboard steel, the desired effect is for the liner to create a larger opening toward the exit of the loading zone to prevent material entrapment.

As noted above, UHMW wear liner is typically installed so its bottom edge will touch or lie on the belt.

At the entry area, the space between the belt and the bottom edge of the wear liner is generally in the 1/8 to 3/8 inch (3 to 10 mm) range, with the closer dimension specified for materials with smaller particle sizes. At the exit end, the distance will typically be 3/8 to 3/4 inch (10 to 20 mm). Again, the smaller distance is for finer materials, while the larger dimension is for materials containing larger lumps. Proper belt support to eliminate belt sag and vibration is essential to preserving the belt in the face of this narrow spacing.

A word of caution: when joining pieces of wear liner, it is imperative the bottom edges line up smoothly without creating a jagged, "saw-tooth" pattern. **(Figure 8.7)** If the bottom edges are not precisely aligned, entrapment points will be created. The conveyed material will then create exceptionally highpressure points in these areas that will lead to material spillage or, even worse, wedges of material will accumulate into "teeth" that will abrade the belt. To prevent these buildups, the bottom edge of the wear liner should be straight, as if a string line were stretched from the entry to the exit of the transfer point. Again, relief from entrapment should be provided with a slight increase in the distance from the belt surface to the bottom edge as the belt moves toward the exit of the transfer point. **(Figure 8.8)**

Methods of Wear Liner Installation

Wear plate can be applied by methods including bolts, welding, or a combination of both.

Wear liners are commonly installed with countersunk bolts that provide smooth surfaces on the interior face of the skirtboard. These bolts also allow the simple replacement of the liner. Liners can be welded into position with the obvious drawback being the difficulty in replacing worn liners. Should the installation require that the wear liner be welded into position, care must be taken to use the correct welding materials and techniques to match the liner material.

Even when the liners are installed by bolts, it may be wise to tack weld the liners so as not to rely totally on the holding power of fasteners that are subject to vibration, abrasion, and impact damage.

Another installation technique calls for the wear liner to be plug welded from outside the transfer point. **(Figure 8.9)** With this technique, holes are drilled or cut through the plate steel wall. Then the back of the liner is welded to the chute wall. This system offers installation without bolt holes or bolt heads protruding into the load zone to act as targets for material abrasion. The liner provides its full thickness for wear life. At the end of the liner's life, replacement can be performed by cutting out the "plug" welds and installing new liners using the same holes.

When welding the liner in place, care must be taken to control the stress introduced into the lining. Abrasion resistant plate, when applied as a liner, must be applied the same way a person would apply wallpaper. If you apply a sheet of wallpaper by securing the four outside edges first, big air bubbles will be trapped in the center of the sheet. You get a similar situation if AR plate is installed in the same fashion. But instead of trapped air bubbles, it is residual stress trapped in the plate that will try to escape. When the structure starts to flex under normal operation, these stresses can introduce cracks in the wear liner. If not caught in time, a large section of liner can break off.

To avoid this stress, it is important to use proper welding technique. The accepted "best practice" is termed back welding. It calls for stitch welding in short pairs, beginning on the top of the plate, and alternating back and forth from the top to bottom. Position each subsequent pair of welds further toward the unwelded end of the liner. At each weld, draw the bead back toward the welded end. **(Figure 8.10)**

Careful attention must be paid to the strength of the conveyor structure when installing liners for the first time. Unless properly reinforced, this structure can sometimes be too weak to support the added weight of liners, risking costly damage and downtime.

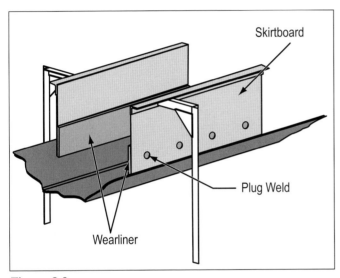

Figure 8.9
Wear liner can be installed from the outside of the skirtboard by plug welding.

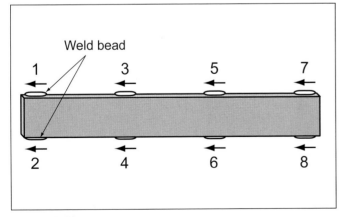

Figure 8.10
Back welding is "best practice" for the installation of wear liner.

Edge Sealing Systems

chapter 9

"The proper application of belt support and wear liner allows the sealing system the opportunity to work effectively."

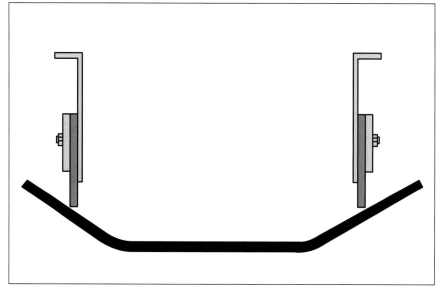

Figure 9.1
The sealing system is installed below the skirtboard to keep the material on the belt.

Another requirement in a transfer point designed for reduced spillage and high efficiency is an effective sealing system at the edges of the belt in the loading zone. **(Figure 9.1)** This is typically a flexible elastomer strip installed on the sides of the skirtboard on either side of the belt to bridge the gap between the steel structures and the moving belt.

Some might argue that the application of the sealing rubber should be the first step or primary requirement for sealing a conveyor transfer point (rather than the final step). After all, the containment of fugitive materials is the sole reason for the sealing system's existence. However, this view does not address the problems of belt sag and the forces of material side loading. The performance of any sealing system suffers when the crucial first steps of installing proper belt support and wear liner are not taken. The proper application of belt support and wear liner allows the sealing system the opportunity to work effectively.

The Goals of a Sealing System

The goals of any sealing system is to effectively contain material fines to keep material on the belt. Desirable attributes include minimum contact area and downward pressure to reduce drag against the conveyor power supply and minimize wear on the belt and seal.

Belts suffer wear along the skirted area at a set distance in from either edge of the belt. Often the sealing system is blamed for this wear. However, most often this wear is not caused by abrasion from the sealing strip, but rather from the material entrapped in the seal or skirtboard seal that is held against the moving belt.

What Sealing Systems Can Do and Cannot Do

In the past, the elastomer sealing strip has been expected to perform an almost miraculous function. It would be installed in a straight line on an unloaded belt, and then be expected to provide a leak-proof seal when the belt was deflected under load. If the deflection (or sag) of the loaded belt could be accurately predicted, and the skirt rubber cut into the proper "roller coaster" shape and installed and maintained in that shape, theoretically an effective seal would be created.

However, as a conveyor belt travels, the seal will wear at different rates along its length, making it unrealistic to attempt to create this type of seal over the long run.

When the belt is not properly supported or there is no wear lining, the elastomer edge seal is asked to act as a dam to contain the full weight of the material load. To ask these flexible sealing strips to do more than contain dust on the belt is asking for the unattainable. Elastomer strips are not suited for this purpose. The pressure from the weight of the material load quickly abrades the belt and the sealing strip, or pushes the strip away from the wall or up from the belt. These failures allow the resumption and acceleration of spillage. **(Figure 9.2)**

In an attempt to stop the escape of particles, plant personnel will adjust their sealing systems down to the belt, increasing the sealing pressure. But this adjustment can lead to several undesirable results. Any pressure beyond a gentle "kiss" of the seal onto the belt results in increased friction. The increase in pressure can raise the conveyor's power requirements, sometimes to the point where it is possible to stop the belt. In addition, the increased friction causes a heat buildup that will soften the sealing strip elastomer. This shortens its life, sometimes to the point where the seal will virtually melt away. Wear will be most obvious at the points where there is the highest pressure. Typically, this is directly above the idlers.

A more practical solution is the application of wear liner and belt support as discussed in this volume. If proper belt support and wear liner are provided, the edge-sealing problem is nearly solved. In that case, the mission of the skirtboard seal becomes only to contain any small pieces of material and fines that pass under the wear liner. This is a chore a well-designed, well-installed, well-maintained sealing system can perform.

Engineered Sealing Systems

The job of sealing a transfer point is challenging. Even in the best transfer points where the belt tracks properly on a stable line, the sealing system must face a certain amount of sideways motion and vibration, due to variations in load and conditions. The sealing system must conform to these fluctuations in belt travel to form an effective seal. The system also needs to be rugged to stand up to belt abrasion and splice impact without undue wear and without catching the splice. The system should also offer a simple mechanism adjustment to compensate for wear over time.

No sealing system can stand up for long in the face of abuse from the material load. If the seals are not sheltered from the material flow, both the effectiveness and the life of the sealing strips will be diminished. When you get the impact of the loading material onto a sealing system, the loading material forces the sealing strips down onto the belt, accelerating wear in both seal and belt. The transfer point should be constructed to avoid both loading impacts on the seals and material flow against the seals.

Figure 9.2
Spillage at transfer points can quickly accumulate in piles beside the conveyor.

Figure 9.3
Laying a large rope against the skirtboard is one crude attempt to prevent spillage.

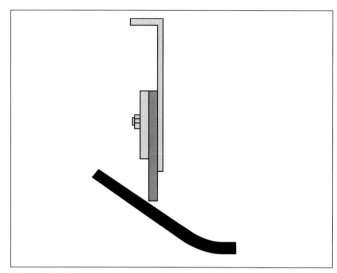

Figure 9.4
"Straight up and down" sealing systems extend down from the skirtboard.

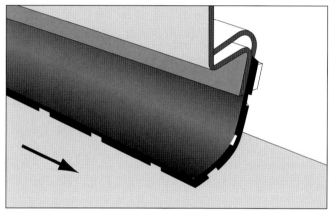

Figure 9.5
A sealing system that lies on the belt underneath or inside the skirtboard may limit the cargo-carrying area.

Figure 9.6
A sealing system that is placed inside the chute wall faces abuse from the material load, and may reduce capacity.

The first skirt seals were fabricated in-house from readily available materials such as used belting and large "barge" ropes. **(Figure 9.3)** These primitive systems were pushed down onto the belt edges and held their position due to gravity. But these systems were not very successful. They could become loaded with material, that would abrade the belt, and they lacked a method to adjust for wear. Eventually, the lack of success with these homemade techniques led to the desire for more effective systems.

In the last fifteen to twenty years, much work has been done to develop systems that effectively meet sealing requirements. The state of the art in the engineering of transfer point components has progressed from sealing strips that could barely contain lumps of material to the current systems that can prevent the escape of fines and even dust. There are a number of sealing systems now commercially available. In general, they consist of long strips of an elastomer held against the lower edge of the skirtboard by an arrangement of clamps.

Innies, Outies, and Straight Up and Down

There are a number of different approaches to skirtboard sealing. Like belly buttons, there are "innies," "outies," and a few that are pretty much straight up and down. In other words, there are some that stick back inside the chute, some that are on the outside of the skirtboard, and some that stick straight down from the skirtboard.

The "straight up and down" systems typically use just a rubber strip; it may be a compliment to call them engineered systems. **(Figure 9.4)** Typically, one supplier offers a system of clamps, and another provides the rubber strip. Sometimes a specially fabricated elastomer strip is installed; other times, strips of used belting have been applied.

A specific caution must be given against the use of old, used, or leftover belting as a skirting seal. Used belting may be impregnated with abrasive materials–sand, cinders, or fines–from its years of service. In addition, all belting contains fabric reinforcement or steel cord. These embedded materials can grind away at the moving belt, wearing its cover and leading to premature failure and costly replacement.

Sealing on the Inside

Some sealing systems are clamped on the outside of the chute, and then curl the sealing strip back under the steel to form a seal on the inside of the chute wall. **(Figure 9.5)** These "innie" systems work well on conveyors with light, fluffy materials and fine, non-abrasive materials such as carbon black.

This may be an attempt to minimize the edge distance requirement and so increase the working capacity of the belt. Another advantage of this design is it is less vulnerable to damage from belt mistracking. But this system has some weaknesses, too.

As noted earlier, wear liner is a major factor in protecting the sealing system. The wear liner prevents side-loading forces of the material from pushing the material out to the edge of the belt. Even the protective benefit from the installation of wear liner is neutralized if the sealing system reaches back under the chute wall and places the sealing strip on the inside of the steel. **(Figure 9.6)** In this position, it can easily be damaged by the load forces.

Other problems have been noted with this "inside the chute" sealing system design. Material can work under the strip as it lies on the belt inside or under the chutework and abrade a wider area of the belt surface. In addition, negative air pressure inside the chute created by the dust collection system may lift these sealing strips off the belt, allowing dust to escape. In contrast, sealing strips that are installed on the outside of the skirtboard steel are pulled down onto the belt by the negative air pressure created by the dust collection system.

In addition to the reduced wear life cited above, this design induces another problem. As this skirting is inside the chute, it can reduce the usable area of the belt by taking up space where the load could be carried. Due to a "rainbow" effect, moving in the edges of the carrying area reduces capacity all the way across the belt. If the skirting seals are positioned inside the chute, the carrying capacity of the belt is reduced. **(Figure 9.7)**

It is important to remember that the chute and transfer point are designed for an established rate of flow by providing a constant load on the belt. But rarely is flow constant. Rather, it is a question of surges, peaks, and valleys. In addition, the action of loading has caused the volume of material to expand beyond its normal density. As a result, the material pile inside the loading zone spreads, so it stands wider on the belt. Due to this "rainbow" effect, moving in the edges of the carrying area reduces capacity all the way across the belt. If the skirting seals are positioned inside the chute, the carrying capacity of the belt is reduced.

The time when the seal should be placed inside the skirtboard is when there is too little free belt area outside the skirtboard to allow effective sealing. In this case, an "outside" design is more likely to create or allow damage to belt or sealing system from belt mistracking and spillage.

Outside Sealing Systems

It is often critical to overall plant efficiency to maintain the capacity of a conveyor. History has demonstrated that demand on systems goes up rather than down. Conveyors are called on to carry more material, at faster speeds, on shorter operating schedules. The last thing any plant should want is to reduce its conveyor capacity by placing its sealing system on the inside of its skirtboard inside the loading zone.

It is much better to place the seals on the outside of the chutework. When the sealing system is positioned outside the skirtboard steel, it is removed and protected from the severe downward and outward forces of material loading. **(Figure 9.8)** The installation of wear liner inside the skirtboard creates a dam to keep the weight of the material load away from the sealing strips. Now the system can seal effectively without reducing capacity and without being subjected to a barrage of

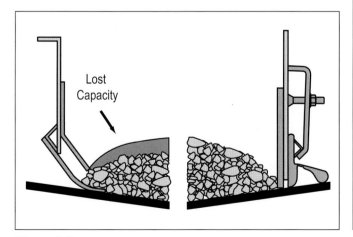

Figure 9.7
Due to the "rainbow effect," moving the edge of the material in will reduce capacity across the full width of the belt.

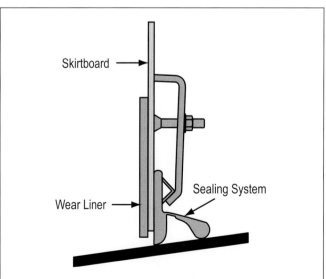

Figure 9.8
Wear liner will protect the sealing system from the force of the material load and allow the seal to contain fines and dust.

material. The conveyor can use the full capacity of the belt without artificial limitations, and the belt's service life is greatly increased.

It is critical that there is adequate free belt area to allow effective sealing. Free belt area–the amount of belt outside the skirtboard on either side of the conveyor–provides the space available for sealing. Too often, in the interest of putting the greatest load on the narrowest belt, the belt free area is reduced. This invariably comes at the price of ineffective sealing systems. The required edge distance for any sealing system must be provided, or the system risks failure.

For more information on free belt area and effective belt width, see Chapter 6.

Dressing in Layers

When outdoor sportsmen such as mountain climbers, hunters, and skiers prepare for cold-weather activities, they dress in layers. They know it is better to put on multiple layers of clothing–undershirt, shirt, sweatshirt, and jacket–than to wear one thick layer. The same is true for transfer point sealing: it is better to work with several thin layers than one thick, all-purpose one.

The first layer is provided by the wear liner installed inside the chute. Extending down to less than one inch (25 mm) off the belt, it keeps the larger particles (and the large volumes of material) well away from the belt edge.

The sealing system provides the next barrier. Now, in the interest of improved performance, sealing systems have been developed that, rather than relying on a single strip, also use multiple layers. These sealing systems incorporate two layers; a primary strip that is pushed gently down to the belt to contain most particles, and a secondary strip that lies on the belt's outer edge to contain any fines that push under the wear liner and primary strip. **(Figure 9.9)** These secondary strips contain a channel to capture the fines, eventually redirecting the particles into the main body of material.

A castle has a moat for outer defense, a wall for an inner defense, and a stronghold or keep for final defense. An effective sealing system uses the same "layer" principles to contain material. The wear liner on the inside of the chute forms the first line of defense, a primary sealing strip forms the inner wall, a secondary seal provides the final barrier to the escape of fugitive material.

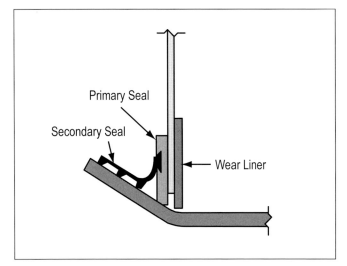

Figure 9.9
An effective multiple-barrier sealing system incorporates wear liner, a primary sealing strip, and a secondary strip.

Figure 9.10
This multiple-layer sealing system uses a primary seal that is pushed gently down on the belt and a secondary strip that captures fines in its ribs.

A Low-Pressure Multiple-Layer System

One successful multiple seal installed on the outside of the skirtboard uses a two-component system. It is composed of two strips of urethane: a primary strip and a secondary strip, installed one in front of the other to form an effective seal. **(Figure 9.10)** The primary strip clamps against the outside of the chutework and extends vertically down to lightly touch the belt. It is applied with light pressure: the clamp applies force horizontally toward the chute, rather than down onto the belt. Because the clamping force is horizontal (against the chute) rather than vertical (down onto the belt), this primary strip contains material without the application of pressure that increases wear and conveyor power consumption. This primary strip contains most of the material that has escaped past the chute's wear liner.

The secondary sealing strip is joined to the back of the primary with a dovetail and lies on the belt's outer edge like an apron. It is designed with ridges or legs molded into its surface to capture any fugitive material that has passed under the primary seal. This fugitive material is then carried down the belt in the tunnels formed by these legs. With these tunnels, the material is trapped until it reaches the end of the sealing strip. By that time, the material has settled, and will fall back into the main material stream.

The secondary sealing strip requires only the force of its own elasticity to provide sealing tension against the belt, and consequently, will wear a long time without the need for adjustment.

This sealing system works well with many types of clamps, allowing reuse of old ones or the purchase of inexpensive new ones. These clamps install on the exterior of the chute or skirtboard without needing any backing plate system.

Because this multiple-layer sealing system lies on the outside of the chute, it does not reduce the conveyor's load capacity and is not subject to abuse from the material load.

Multiple Layers with a Single Strip

Recently, a twist has been added to this multiple-layer concept. New designs combine multiple-layer seal effectiveness with the simplicity of single strip systems. These designs feature a rugged single-strip that is manufactured with a molded-in flap that serves as a secondary seal. **(Figure 9.11)** This "outrigger" contains the fines that have passed under the primary portion of the single strip. As this sealing strip is a one-piece extrusion, installation is a simple procedure without requiring the assembly of two components. Single-piece, multiple-layer sealing systems are now available in both standard and heavy-duty constructions.

Floating Skirting Systems

Another sealing system uses sealing strips mounted to the steel skirtboard on independent, freely-rotating link arms. **(Figure 9.12)** The links allow the sealing strip to float, reacting with changes in the belt line while remaining in sealing contact with the belt. This system also allows the sealing system to self-adjust, using its own weight to compensate for wear. **(Figure 9.13)**

Edge Distance Requirements

Care must be taken in selecting an edge sealing system to fit into the available distance between the belt's edge and the steel skirtboard. It is generally not a good idea to have the sealing strips all the way out to the belt edge, as this risks damage in the event of even minor belt mistracking. It is better to have additional belt width outside the sealing strip to allow for belt wander and to act as an additional distance for material to pass before escape.

In general, three inches (75 mm) of edge distance outside the chute wall is recommended as the minimum distance to establish an effective seal for conveyor belts 18-30 inches (450 to 750 mm) in width. For conveyor belts 36 inches (900 mm) and above, a minimum of four inches (100 mm) edge distance is recommended. As the edge distance is diminished toward the minimum, it becomes more critical to control belt tracking through the use of belt training devices. *A more detailed discussion of skirtboard width and belt edge distance is presented in Chapter 7: Skirtboard.*

Figure 9.11

This multiple-layer sealing system is composed of a single strip of elastomer.

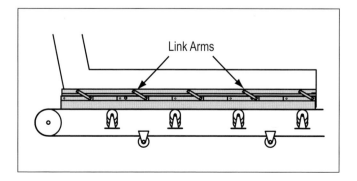

Figure 9.12

The floating skirting system remains in contact with the belt as its line fluctuates under various loading conditions.

Power Requirements of Sealing Systems

In order to be effective in keeping material on the belt, sealing materials must exert some amount of pressure against the belt. This pressure will increase the drag against the belt, and hence increase the power consumption of the conveyor. This additional power requirement is generally independent of the width of the belt or the width of the seal material. It is directly dependent on the length of the seal and the pressure applied to the seal to keep it down on the belt.

In the book *Belt Conveyors for Bulk Materials*, CEMA uses a pressure of three pounds tension per foot (4.5 kg/m) of seal as its standard reference. Tension equals the force perpendicular to the belt times the coefficient of friction. (Remember, any transfer point must be sealed on both sides, effectively doubling the tension added per unit length.) In test facilities and actual field situations, it has been found that many skirtboard sealing systems can be adjusted down onto the belt with very high levels of force. In these cases, the CEMA standard underestimates the actual tension applied to the belt.

The multiple-layer sealing system recommended above relies only on the pressure supplied by the resilience of the "outrigger" leg, and therefore seals with much less pressure down against the belt. The clamping pressure is applied horizontally against the steel skirtboard rather than vertically against the belt. It applies much less tension to the belt, and consequently consumes less conveyor power.

For demonstration purposes, the pressure that sealing materials exert against the belt was chosen as 6 lb/ft (8.9 kg/m) for rubber strip, rubber segments, and UHMW wear liner/seal combination, and 1 lb/ft (1.5 kg/m) for the multiple-layer seal. These figures have been found to present a reasonable approximation of the start-up power requirements as measured in the field. The running power requirement is typically one-half to two-thirds of the start-up power requirement. If actual conditions are known, the actual power requirements or tension should be calculated and used.

Comparing Sealing System Power Consumption

Figure 9.14 displays the approximate power consumption for sealing systems composed of the following materials:
- A. Rubber Slab (60 to 70 Shore D SBR Rubber).
- B. UHMW Wear Liner/Seal Combination (Wear liner extends down to touch the belt).
- C. Multiple-Layer Sealing System (Urethane).
- D. CEMA Standard (For Reference).

This chart is based on friction factors of 0.75 for a rubber sealing system on rubber belt and 0.545 for UHMW side rail support on rubber belt. These friction factors are based on actual tests against rubber belting, both with and without material present.

Figure 9.13

Link arms allow this sealing system to self-adjust, using its own weight to compensate for wear.

These tables provide a means of estimating the power required for edge sealing. By multiplying the length of one side of a transfer point by the belt speed and then studying the figures, a reasonable estimation of start-up power consumption required by the conveyor sealing system can be determined.

The power requirements for Sealing Systems A., B., and C. above have been calculated using a side rail sealing support cradle under the belt. The CEMA standard was determined using standard idler belt support.

For example, a transfer point that is 20 feet (6 m) long and operates at 400 fpm (2 m/sec) with a rubber strip seal and side rail support would have a start-up power requirement of approximately 3.7 hp (2.8 kW). In contrast, the use of a multiple-layer sealing system on the same transfer point would consume only 0.62 hp (0.5 kW).

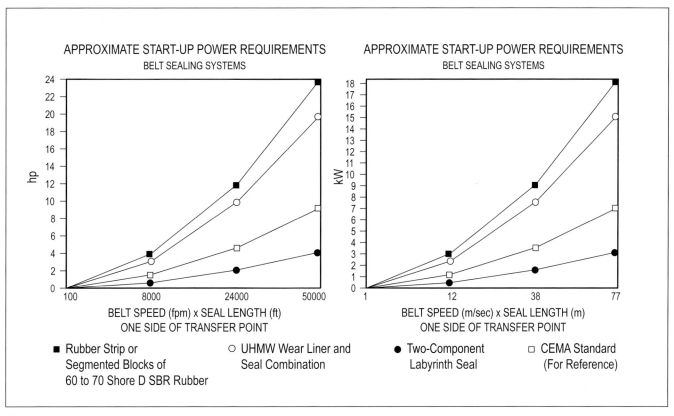

Figure 9.14
Approximate Start-Up Power Requirements of Various Belt Sealing Systems.

Sealing Systems and Belt Cover Wear

A research project was done to determine to what extent engineered belt cleaning and sealing systems increased or decreased belt wear. This study tested the abrasion of several edge sealing systems against a typical conveyor belting. The conclusions of the study report the use of more sophisticated belt cleaning and sealing systems with adequate maintenance can extend the life of the conveyor belt. While belt wear is introduced by these devices, the amount of this wear is approximately one-half the rate to be expected when the belt runs through accumulations of fugitive material, resulting from the lack of (or failure of) cleaning and sealing systems.

Skirt Seals and Belt Wander

All skirt sealing systems are vulnerable to damage from a mistracking belt. If the belt wanders out from underneath one side of the skirtboard, the then-unsupported sealing strip will hang down below the line of the belt, regardless of whether its normal position is inside or outside the chute. When the belt moves back into the proper position, the seal will be abraded from contact with the edge of the moving belt. Or it can be bent backward into an unnatural position and worn away. Either outcome risks a significant increase in spillage.

Having adequate edge distance is an advantage, as the more free belt area provided, the more the belt can wander before it risks significant damage. But key to avoiding this damage, of course, is to control belt tracking.

For information on control of conveyor belt tracking, see Chapter 15.

Avoiding Grooves in the Belt

It is a common misconception that the sealing system must be "softer" than the belt cover, so the seal wears before the belt. Usually softer materials are more cut resistant and, therefore, provide longer life. However, it is possible to make seals from materials with a wide range of hardness and wear resistance. The key is to avoid material entrapment and high pressure against the belt.

Grooves in the belt underneath the skirtboard are often blamed on the sealing system. These grooves can be caused by excessive pressure between the skirt and the belt that can heat the cover to the point it loses its wear resistance.

Grooves are often started by the entrapment of lumps of material between the liner and the belt. This starts out scratching the surface and gradually wears away the belt cover. Loading while the belt is in transition makes it easy for material to be trapped under the skirtboard.

A key to minimizing grooves in the belt is to make the skirt sealing system sacrificial. This is done by making sure the sealing strip has a lower abrasion resistance than the belt and by making sure the strip has the flexibility to move out of the way when a lump of material moves underneath the skirtboard.

To avoid grooving the belt, a transfer point must have proper belt support, effective wear liner, and light sealing pressure.

Installation Guidelines

A sealing system must form a continuous unit along the sides of the steel skirtboard. If simple, end-to-end butt joints are employed to splice lengths of sealing strip together, material will eventually push between the adjoining surfaces and leak out. An interlocking or overlapping joint is best to prevent this spillage.

With all strip skirting systems, it is a good idea to "round off" the strip's leading edge at the tail end of the conveyor where the belt enters the back of the loading zone. **(Figure 9.15)** By presenting a rounded edge to the belt, this reduces the chance a mechanical belt fastener can catch the strip to rip it or pull it off the chute.

Maintenance of the Sealing System

When specifying a skirt sealing system, it is wise to consider the mechanism for the adjustment and replacement of the wearable rubber. As the conveyor runs, the heat generated by the friction of the belt against the skirting seal combines with the wearing action of the fines to erode the sealing strip. To counter this wear, the sealing strip must be adjusted down against the belt.

Elastomer skirtboard edging should be adjusted so that the edging just "kisses" the belt surface. Forcing the edging hard against the belt cover will lead to extra wear in both the belt and the seal because of the additional heat generated from the increased friction. Applying too much pressure to the sealing system will also require additional power to move the belt. On conveyors with lengthy skirtboards, rubber sealing systems applied with excessive pressure may overload the conveyor's drive motor, particularly at conveyor start.

However, if the procedures for service of skirting rubber are cumbersome or complicated, three detrimental consequences are likely:
1. Adjustment does not happen at all, so the skirting sealing strip wears, and leakage resumes.
2. Adjustment is made only rarely, so spillage occurs intermittently.
3. The maintenance man or conveyor operator, compensating for not making regular adjustments, will over-adjust the rubber, forcing it too hard and too far down onto the belt. This risks damaging the belt or catching a splice and ripping out the entire section of skirting.

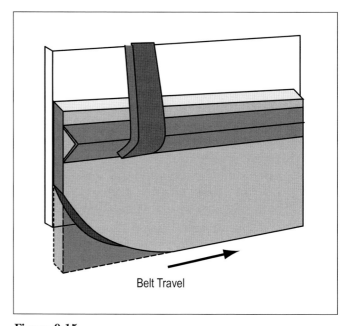

Figure 9.15
Rounding the leading edge of the sealing strip will avoid damage from mechanical belt splices.

To prevent these problems, maintenance procedures should be as free of complications, tools, and downtime as possible. The simpler it is to adjust the sealing strip downward onto the belt–the fewer fasteners to loosen, the fewer clamps to remove–the more likely it is that the adjustments required to maintain an effectively sealed, low-spillage transfer point will actually take place.

Skirting systems that lie on the belt in some fashion can minimize the need for maintenance adjustment. The forces of gravity combine with the resilience of the rubber to keep an effective sealing posture.

The multiple-layer sealing system provides a self-adjusting function, where elastomeric memory maintains the sealing pressure. As the "legs" of the secondary strip wear, the internal resilience of the strip keeps it down onto the belt, maintaining seal effectiveness. The primary strip may require periodic adjustment downward to allow for wear and belt pressure upwards. This adjustment can be accomplished by loosening the clamps and pushing the strip down. It can even be performed with a judicious hammer blow onto the top of the primary strip.

Choosing a Skirtboard Sealing System

When selecting a skirtboard sealing system, it is important to match the severity of your application. Factors such as belt speed, material load, and free belt area should be reviewed to make sure the application receives a suitable system.

To prevent the sealing strips from wearing the belt, the sealing strips should be composed of material that has a lower abrasion resistance than the top cover. That way the seal will wear before the belt cover. Note that abrasion resistance is not measured by durometer (which is a rating of hardness) but rather an abrasion index, such as Pico, DIN, or Taber ratings. ASTM D2228 recognizes the Pico abrader as the standard test method.

It is advisable to specify a multiple-barrier sealing system, as this will afford more opportunities to contain the material. The barriers should includes legs or grooves to capture fine particles and allow the particles to continue along on the conveyor until they can eventually return to the main material body.

You are well advised to install long strips that will allow a continuous seal along the load zone. This will avoid the butt joints that are typically the first place for sealing problems to appear.

Periodic inspection and maintenance will prevent damage, extend life, improve performance, and boost satisfaction. That will ensure that you receive optimum value for your investment in an engineered sealing system.

The Final Step In Spillage Control

Rather than being the first step in solving conveyor spillage, the skirtboard seal is the last chance to corral fugitive material and prevent its release. The better the job done with the belt support and wear liner systems to contain material and keep it away from the edge, the better will be the performance from the belt's edge sealing system. A multiple-layer system will provide effective material containment for a transfer point and improve the operations of the belt conveyor.

Chapter 10

Control of Air Movement

Conveyor loading zones and discharge points are prime sources for the creation and release of airborne dust. Depending on a number of factors, including the nature of the material carried, the height of drop onto the belt, and the speeds and angles of unloading and loading belts, various systems to control airborne dust should be installed at conveyor loading and unloading zones.

The Basics of Dust Control

The conditions that determine whether fine materials become airborne are air velocity, particle size, and cohesion of the bulk material.

The relationship of these characteristics can be summarized as:

$$\frac{\text{Air Velocity}}{\text{Particle Size x Cohesiveness}}$$

If air velocity is increased, but particle size and cohesiveness remain constant, then dust will increase. If air velocity remains constant and particle size and/or cohesiveness are increased, the amount of airborne dust will be reduced.

Where one or more of these parameters is a given, the ability to control dust depends on altering one or both of the other characteristics. For example, where the size of coal particles being transported cannot be changed, the air velocity or cohesive force of the particles must be altered to minimize dust emission.

Whenever a dry material is moved or changes its direction in a process, the result is airborne dust. Dust emissions can be significantly reduced through the use of an engineered sealing system, the addition of a dust suppression system, and/or the use of an effective dust collection system.

Minimizing Dust at Transfer Points

The first consideration in dust control should always be the minimization of the amount of dust actually created. While it is unlikely dust can be completely eliminated, any change in system design or production technique that will reduce the amount of dust produced should be considered.

For example, if the energy released by the falling stream of material at the impact area can be reduced, then less energy will be imparted to the material, and the fewer dust particles created or driven off. Consequently, it is best to design conveyor systems with minimal material drop distances.

> "Whenever a dry material is moved or changes its direction in a process, the result is airborne dust."

This reduces the amount of energy imparted to the fines, and cuts the amount of dust driven off into the air. Other measures can be undertaken to control the amount of dust created and the force of air currents that can pick up particles and carry them outside the loading zone.

Since it is generally not possible to prevent the creation of dust, systems to suppress and capture it must be employed. In their simplest form, these dust control involve nothing more than attention to this requirement during the engineering of the transfer point chutework. More sophisticated approaches include the addition of moisture to return the dust to the main material body and/or the use of filtration systems to collect dust out of the air.

Air Movement Through Transfer Points

The size and cost of the transfer point enclosure and the other components of a conveyor's dust management system are directly related to the volume of air that must be pulled through the system. An understanding of air movement through the transfer point will allow the designer much greater success in controlling the release of fugitive material. The understanding and control of air movement is, therefore, fundamental to efficient and economical dust control.

Ideally, a slight negative pressure is wanted inside the enclosure. This condition would pull air into the enclosure so that fines and airborne dust are retained in the structure, rather than carried out. But typically, this is difficult, if not impossible, without a large dust collection system. The airflow created by the equipment above the transfer point and the movement of material through the transfer point generate a positive pressure through the system, creating an outward flow of air. If this positive pressure is not addressed with adequate pressure relief or dust collection systems, the particles of dust will be carried out of the transfer point on the outward airflow.

For adequate control of dust, the collection system's exhaust air volume must be equal to or greater than the rate of air movement.

Measurement of Air Quantities

Many companies and consultants have developed formulas for the calculation of the amount of air moving through process systems such as transfer points. The following is offered as a relatively simple, theoretical, yet workable method. Of course the conditions of any specific combination of conveyor and material will affect the results significantly.

There are three sources of air movement that might be present in a given transfer point. They are Displaced Air, Induced Air, and Generated Air.

The total air flow in a given transfer point can be determined with the following equation:

$$Q_{tot} = Q_{dis} + Q_{ind} + Q_{gen}$$

			Measured in	
			Imperial	Metric
Where: Q_{tot}	=	total air movement	cfm	m³/sec
Q_{dis}	=	calculated displaced air	cfm	m³/sec
Q_{ind}	=	calculated induced air	cfm	m³/sec
Q_{gen}	=	determined generated air	cfm	m³/sec

Displaced Air

A good example for displaced air is a coffee cup. When coffee is poured into the cup, the air inside is displaced by the coffee. This same effect occurs when material enters a load chute: the air that filled the chute is pushed out, displaced by the material. This amount of air displaced from the chute is equal to the volume of material placed into the chute.

To calculate Displaced Air: (cfm)

$$Q_{dis} = \frac{\text{pounds per minute (of the material conveyed)}}{\text{bulk density (lb/ft}^3\text{)}}$$

$$\text{pounds / minute} = \frac{\text{tons per hour} \times 2000}{60}$$

To calculate Displaced Air: (m³/sec)

$$Q_{dis} = \frac{\text{kg/second (of the material conveyed)}}{\text{bulk density (kg/meter}^3\text{)}}$$

$$\text{kg / second} = \frac{\text{metric tons per hour} \times 1000}{3600}$$

Induced Air

Induced air is air collected by the moving product as it leaves the head pulley. As it lies on a belt, conveyed material has a certain amount of air entrapped in the product. However, as the material leaves the head pulley in a normal trajectory, the body of product becomes larger. Each particle of material collects and gives energy to an amount of air. When the product lands and compresses back into a pile, this induced air is released, causing substantial positive pressure flowing away from the center of the load zone. If this positive pressure is not addressed with proper relief systems, the dust fines will be carried out of the system on this air current.

This equation provides an approximation of the amount of induced air. The recommended exhaust volume should be equal to or greater than the total calculated volume.

The most controllable, and hence most important factor is A_u, the size of the opening through which the air induction occurs. The smaller the opening(s) for air

The reference *Dust Control Handbook* offers the following formula to calculate the amount of induced air:

$$Q_{ind} = 10 \times A_u \times \sqrt[3]{\frac{RS^2}{D}}$$

		As measured in	
		Imperial	Metric
Q_{ind}	= volume of induced air	cfm	m³/sec
A_u	= open area at upstream end at the point where air is induced into the system by action of falling material	square feet	square meters
R	= rate of material flow	tons / hour	metric tons / hour
S	= height of material free fall	feet	meters
D	= average material diameter	feet	meters

Type Of Crusher	At Crusher Feed per foot (300 mm) opening width	At Crusher Discharge per foot (300 mm) conveyor width
Jaw Crusher	500 cfm (850 m³/hr)	500 cfm (850 m³/hr)
Gyratory Crusher	500 cfm (850 m³/hr)	1000 cfm (1,700 m³/hr)
Hammermills and Impactors	500 cfm (850 m³/hr)	1500 cfm (2,550 m³/hr)

Figure 10.1
Approximate Levels of Generated Air from Various Types of Crushers.

to enter the system, the smaller the value of A_u and the smaller the volume of exhaust air. In other words, if the door that the air comes in is small, there will be less air to exhaust. Note: This is the size of the entry upstream rather than the size of the loading zone or receiving conveyor.

Generated Air

Other sources of moving air are the devices that feed the conveyor load zone. These would include equipment such as crushers, wood chippers, hammer mills, or any device with a turning motion that creates a fan-like effect. While not present in all transfer points, this air flow, called Generated Air, is in many instances the most severe of all air movements.

This type of air movement can be measured using air velocity and volume gauges, such as Pitot tubes and manometers. But it might be simpler for the end-user to contact the equipment's manufacturer to obtain a calculation of Generated Air. **Figure 10.1** displays one crusher manufacturer's approximate figures for the air generated by various types of equipment.

Air Velocity

Air flows from a high- to a low-pressure zone because of the pressure difference. The speed and the volume of air and the speed of air movement are related according to the following equation.

$$Q_{tot} = A \times V$$

			Measured in Imperial	Metric
where	Q_{tot}	= Volume of airflow	cfm	m³/sec
	V	= Velocity of air	FPM	m/sec
	A	= Cross Sectional Area through which air flows	feet²	m²

The amount of air and the size of the area through which it moves determine the speed at which this air moves. To determine the air velocity, this formula is used:

$$V = \frac{F}{A}$$

			Imperial	Metric
where:	V =	Velocity	fpm	m/sec
	F =	Flow	cfm	m³/sec
	A =	Area	feet²	m²

When calculating the cross-sectional area of the exit chute, the area occupied by the body of material must be subtracted. The measurement is for air space only.

A good design parameter for the sizing of the exit chutes of load zones to enhance dust collection and dust suppression is to maintain in the exit area an air velocity below 250 feet per minute (1.25 meters per second). Higher velocities of air movement will allow the air current to pick up particles of material and hold them in suspension in the air, making collection or suppression more difficult.

Controlling Air Entrainment

A complete system to control dust at conveyor transfer points is based upon three design parameters:
- To limit air coming into the enclosure and the material stream.

- To reduce the creation of dust inside the enclosure.
- To lower air velocities within the enclosure, allowing suspended dust to fall back on the conveyor belt.

Air is barred from entering the enclosure at the head pulley of the discharging conveyor by conventional rubber curtain seals at the belt's entrance to the enclosure and by keeping the material in a consolidated stream as it moves through the transfer point. It is also important to keep inspection doors and other openings sealed while the belt is in operation.

As it moves through the transfer point, each particle or lump of material acts on the air in the enclosure, carrying some of the air along with it. In a conventional conveyor discharge, the material free-falls. This disperses the material, making the stream larger and able to take more air with it as air fills the voids created within the spreading material. When the material lands, the entrained air is pushed away from the pile, creating a positive pressure.

The "Hood and Spoon" System

Keeping the material in a consolidated body will reduce the amount of air that will be entrained or carried along. A "hood and spoon" design is used to confine the stream of moving material. **(Figure 10.2)** The hood minimizes the expansion of the material body, deflecting the stream downward. The spoon provides a curved loading chute that provides a smooth line of descent so the material slides down to the receptacle–whether that is a vessel or the loading zone of another conveyor. The spoon "feeds" the material evenly and consistently, controlling the speed, direction, and impact level of the material in the load zone.

By reducing the velocity and force of material impact in the load zone to approximate the belt speed and direction, this system mitigates splash when material hits the receiving conveyor. Therefore, there is less dust and high velocity air escaping. As the material is deposited more-or-less gently on the belt, there is minimal tumbling or turbulence of the material on the belt. There is less impact, which will reduce damage to the belt, and less side forces that push material out the sides of the belt.

In some cases either the hood or spoon is used, but not both. For very adhesive materials, the hood can be used to direct the stream downward for center loading. This variation is often seen on overland conveyors handling highly variable materials or when handling sticky materials like nickel concentrate or bauxite. Gravity and the flow of material will tend to keep the hood from building up and plugging the chute. In other cases with free-flowing materials, the spoon only is used to change the direction of the stream to minimize belt abrasion and skirt side pressure. Spoons are prone to backing up or flushing if the characteristics of the bulk solids are variable. Some compensation can be designed into the spoon for variability of materials. Sometimes there is not enough space to include both in the design.

The main drawback to using the "hood and spoon" concept is the added maintenance costs for these specially-designed components. Even so, where they can be applied and maintained, they will offer significant benefits in reduced dust, spillage, and belt wear.

This "hood and spoon" system works best when the material stream is kept as close as possible to constant flow. Peaks and valleys in the loading pattern will conflict with the desire to provide a consolidated material stream. The success of this system may well eliminate the need for dust collection systems–that is, baghouses–in many operations.

Exit Area Dust Curtains

The final phase of this "hood and spoon" system is to

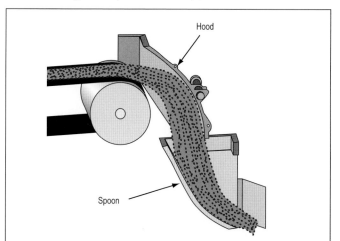

Figure 10.2

By confining the material stream, a "hood and spoon" chute minimizes the air entrained with the material, and so reduces airborne dust.

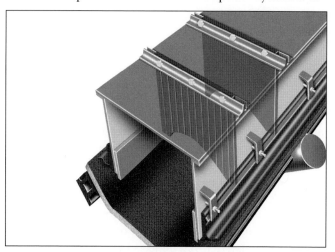

Figure 10.3

Dual dust curtains slow air movement, allowing dust particles to drop out of the air.

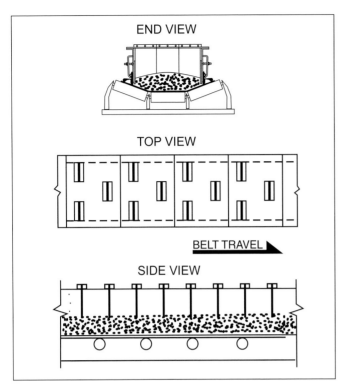

Figure 10.4

Dust curtains installed in a "staggered" pattern form a stilling zone to help control dust.

Figure 10.5

To minimize the chance that a dust curtain will push material off the belt, the curtain should be slit at intervals across its width.

lower the air speed in the enclosure. The goal is to quiet air velocities in order to let dust settle and fall back onto the belt. The way to accomplish this is twofold: make the enclosure large enough and then slow the air movement. By slowing the flow and containing the air, more material can now fall out of the stream within the confines of the enclosure. These baffles are typically rubber dust curtains hanging from the top of the covered skirtboard. **(Figure 10.3)**

These rubber exit curtains inhibit the amount and velocity of air passing out of the loading zone, and so minimize the escape of dust. In most cases at least two curtains should be installed. Some installations –especially those where dust collection and/or suppression systems need to be isolated– may require more.

These rubber curtains can be fabricated as wide as the skirtboard, or one-third its width, and installed in alternating or "staggered" fashion. **(Figure 10.4)** The curtains form a "stilling zone" to reduce airflow and allow dust to settle out.

The curtains should be composed of 60 to 70 durometer. These curtains should be composed of 60 to 70 durometer elastomer and extend to roughly one inch (25 mm) below the top of the pile of conveyed product. These form a barrier that deflects when material is running, yet closes down the opening when there is a void in the material load. The curtain rubber should be slit at regular intervals across the belt width to minimize the chance the curtain will push conveyed material from the product stream and off the belt. **(Figure 10.5)**

Rather than place the curtains at the end of the covered chutework, it is better they be installed inside the covered skirtboard. When the curtain is at the end of the steel enclosure, any material particles hit by the curtain can be displaced from the belt. By placing the curtains so the final curtain is one belt width from the end of the enclosure, any material that contacts the curtains still has room to settle into a stable profile within the confines of the enclosed area. The curtains should be hung roughly 18 inches (450 mm) apart, forming a "dead" area where dust can settle, or dust collection or suppression systems can be applied.* If these curtains are used to isolate dust suppression and/or dust collection installations, it is better if they be positioned the length of one standard chute cover–36 inches (900 mm) apart.

The curtains should allow easy maintenance access to the chute and should be readily removable to allow replacement in the event of wear.

Sealing at the Entry Area

Another technique employed to minimize induced air is to cover the inbound portion of the conveyor for several feet before it enters the head chute. The inbound enclosure is a mirror of the enclosure used in the load zone, complete with skirtboard seals and dust curtains. The air-supported conveyor is particularly well suited for this technique, as it does not require skirt seals.

*Use of dual dust curtains in combination with dust suppresion systems is a patented technology of Raring Corporation.

Control of Air Movement

One technique to reduce air induction at the belt entry is the installation of a piece of used belting between the return run and the carrying run. Placed across from one chute wall to the other, the belting acts as a wall, enclosing the head pulley and reducing air movement.

Filter Relief Systems

It is important that positive air pressure–the force of air moving through the loading zone–be relieved in a manner that will minimize the pressure against the sealing system and so reduce the release of dust. As noted above, where the volume of air is high–greater than 1000 cfm (0.5 m³/sec)–a mechanical dust collection system should be installed. Where air movement in the load zone is less than 1000 cfm (0.5 m³/sec), the load zone can be fitted with a passive system to filter particles from the air. A passive approach is the installation of at least one filter relief bag.

Basically, these systems consist of an open door with a filter bag, sock, or sleeve stretched over the top, through which dust-bearing air passes out of the transfer point. The bag filters the dust from the air. A transfer point may require installation of more than one of these systems, depending on the size of the bag and the flow of air. These can be attached with a simple circular clamp to the chute wall or skirtboard cover. **(Figure 10.6)**

A typical filter sleeve is made of a durable filter fabric that will capture the dust, yet release the material when desired. **(Figure 10.7)** The dust can be released by mechanical or manual shaking, or even by the bag partially collapsing when the outflow of air stops during conveyor downtime. To reduce the risk of explosions, these relief bags must be fabricated with grounding straps to disperse static electricity. The bags usually contain a grommet, allowing them to be hung from overhead supports. In installations where these pressure relief bags are subject to environmental influences, such as snow or rain, the bags should be installed in a protective shelter.

Is Mechanical Dust Collection Required?

If the positive air pressures are not addressed with proper relief mechanisms or a dust collection system of the correct size, there will be a significant outflow of air that will carry particles of dust.

A mechanical dust collection system will be required when the total airflow in the loading zone exceeds 1000 cfm (0.5 m³/sec), or if the air velocity in the enclosure cannot be lowered to less than 250 feet per minute (1.25 m/sec). The total air movement should be overcome by the addition of mechanical dust collection equipment with a capacity equal to or greater than the total airflow.

Details on the location of dust pickups and amount of air to be removed through mechanical dust collection systems are included in Chapter 12: Dust Collection.

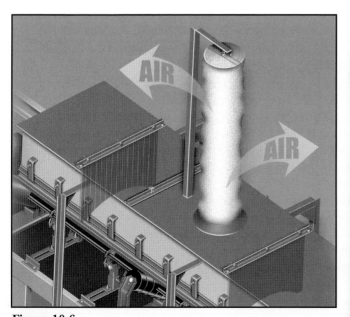

Figure 10.6
By capturing dust while allowing air to pass through, a filter bag serves as a passive dust collection system.

Figure 10.7
A filter bag is installed in the roof of a transfer point enclosure.

Chapter 11

Dust Suppression

"The suppressed dust returns to the main body of conveyed material and on into the process, without requiring additional material handling equipment."

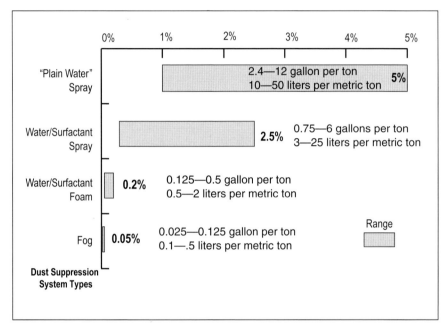

Figure 11.1

Typical Rates of Moisture Addition for Dust Suppression Systems.

Dust suppression is the application of water and/or chemicals either to the body of material, to prevent fines from being carried off into the air, or to the air above the material, to return airborne fines to the material bed.

One big advantage of dust suppression is that the material does not have to be handled again. The suppressed dust returns to the main body of conveyed material and on into the process, without requiring additional material handling equipment.

There are a number of systems used for this purpose, ranging from "garden hose" technology up through water and surfactant sprays, foam, and fog generation systems. These various suppression technologies call for adding different amounts of moisture to the material. (**Figure 11.1**) presents typical amounts of added moisture.

Water Suppression

Perhaps the oldest method for controlling fugitive dust is the application of water over the body of material. By wetting the fines, either as they lay in the material body or as they are being picked up into the air, the weight of each dust particle is increased so it is less likely to become airborne. The moisture also increases the cohesive force of the material body itself, creating larger, heavier groups of particles, and making it more difficult for air movement to carry off fines. This can be done by applying the water through a series of properly sized spray nozzles at a point where the material expands and takes in air, such as during discharge from the head pulley in a transfer chute.

Water can also be applied to create a "curtain" around a transfer point, so any dust fines that become airborne come into contact with the water sprays surrounding the open area. The water droplets are expected to make contact with the dust fines, increasing their mass to remove them from the air stream.

The most effective sprays are low-velocity systems. High-velocity sprays can add energy and a corresponding increase in velocity to the air and the dust particles. This energy is counterproductive to the task of keeping (or returning) the dust with the material body. The high-velocity air movements can keep dust particles in suspension.

Water-based suppression systems can become more sophisticated as the engineering moves beyond "garden hose" technology in efforts to improve results. The effectiveness of water spray systems is dependent on the velocity of applied water, the size of nozzle opening, and the location of the spray nozzles.

Techniques to improve water-spray dust suppression include a reduction of droplet size, an increase in droplet frequency, an increase of the droplet's velocity, or a decrease in the droplet's surface tension, making it easier to merge with dust particles.

Plain water spray application systems are relatively simple to design and operate, but water has only a minimal residual effect, as once the water evaporates, the dust suppression effect is gone. Water is generally inexpensive, it is usually easy to obtain, and it is safe for the environment and for workers who come into contact with it.

Dust suppression systems utilizing water are relatively simple systems that do not require the use of costly, elaborate enclosures or hoods. They are typically cheaper to install and use far less space than dry collection systems. Changes can be made after start up with minimum expense and downtime. Unfortunately, the application of water has several liabilities as well.

Large droplets of water are not very good at attaching to small dust particles due to their associated aerodynamics, high surface tension, and large size. So to increase the result, more water is applied, which is not necessarily a good thing.

With Water, Less Is More

A simple, plain water spray may *appear* to be the most inexpensive form of dust control available. The water is available almost free in many operations (such as mines), and it can be applied through low-technology systems. But this cost justification can be a false equation. Many bulk solids are hydrophobic; they have a high surface tension and are averse to combining with water. In an effort to achieve effective suppression, the amount of water is increased, but because the material does not mix well with water, there will be some particles that remain dry and others that become very wet. This can lead to problems with handling the material, which now accumulates on chute walls, plugs screens, and carries back on conveyor belts.

When applying water to conveyor systems, a good axiom is "less is more." The addition of moisture can cause material to stick together, complicating the flow characteristics of the material being conveyed. For mineral handling in general, the addition of excess moisture prior to screening can cause material to adhere to screen cloth, blinding the equipment. Excess water may promote belt slippage and increase the possibility of wet (and hence sticky) fines accumulating within chutes and around the transfer point.

Problems occurring in "plain water" dust suppression systems include the possibility of excessive moisture in the material, which can downgrade performance in power generation or other thermal processing. Specifically, excess water addition to coal and coke used for boiler fuel results in a BTU penalty that can have a detrimental effect on utility heat rates. The more water added, the greater this penalty.

The Thermal Penalty for Added Moisture

There is a substantial performance penalty added to combustion and other thermal processes when the water content of the fuel is increased. **(Figure 11.2)** In applications like coal-fired power plants and cement plants, water added to the material going into the thermal process must be "burned off" by the process. This can dramatically reduce the process efficiency and increase fuel costs.

It requires 1,320 BTU per pound (3,064 kilojoules per liter) to raise water from 70° F (21° C) to its vaporization temperature of 300° F (149° C). It only takes 20 pounds (9.1 kg or 9.1 liters) of water to increase the moisture content of one ton of material by one percent. As a gallon of water weighs approximately 8.3 pounds (3.8 kg), the addition of less than 2.5 gallons (9.1 liters) of water to a ton of material will raise the moisture content of a ton of material by one percent. Vaporizing just this modest amount of water produces a heat loss of 26,400 BTU (27852 kilojoules).

Because a "plain" water spray requires the highest volume of moisture for effective dust suppression, this method extracts the highest thermal penalty. While the

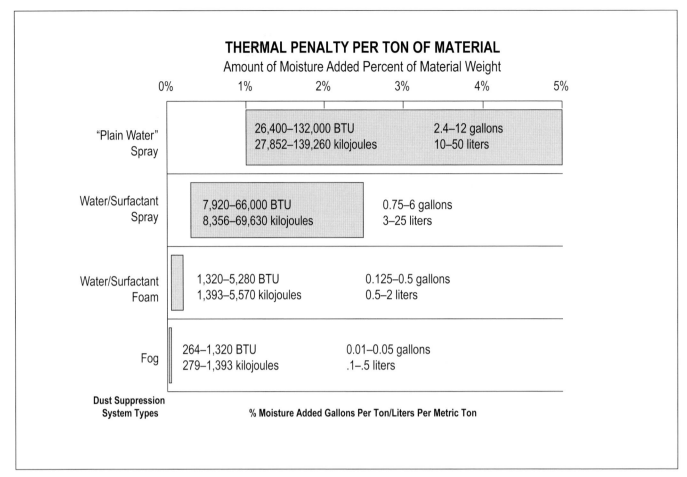

Figure 11.2
Thermal Penalty from Moisture Addition.

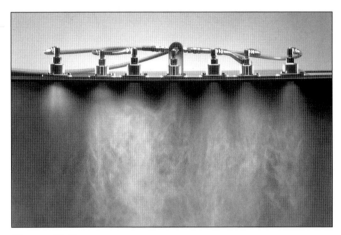

Figure 11.3
Fog dust suppression systems use special nozzles to create a mist of fine droplets.

use of a simple water spray for dust suppression may be the lowest cost, because the water is readily available and there is less "out-of-pocket" expense; the penalty for the addition of surplus moisture can be very costly indeed. Methods to improve dust suppression, while limiting the addition of water, include the use of a "dry fog" or the addition of surfactant chemicals to water that is then applied as a spray or a foam.

Fog Suppression Systems

Fog suppression is one method to optimize the application of water to dusty materials. These systems use special nozzles to produce extremely small water droplets in a dispersed mist. **(Figure 11.3)** These droplets mix and agglomerate with dust particles of similar size, with the resulting larger combined particles falling back to the material body.

The fog systems are based on the knowledge that a wet suppression system's water droplets must be kept within a specific size range to effectively control dust. If the water droplets are too large, the smaller dust particles typically just "slipstream" around them, pushed aside by the air around the droplets. **(Figure 11.4)**

The fog system atomizes the water, breaking it down into very fine particles that readily join with the similarly sized particles of airborne dust. The increased weight of combined particles allows them to settle back to the main material stream. Fog systems supply ultra-fine droplets that maximize the capture potential of the

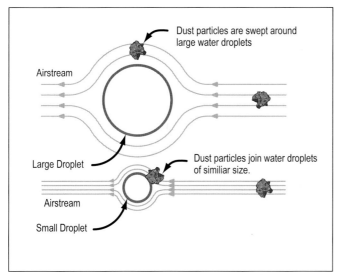

Figure 11.4
Dust particles may miss larger water droplets but join readily with droplets of similar size.

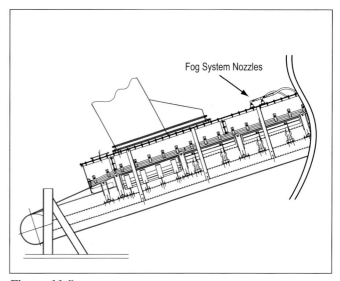

Figure 11.5
Fog suppression systems as applied near the end of the transfer point enclosure.

water while minimizing the amount of water added to the product.

Atomization is designed to reduce the surface tension of the water droplets, while increasing the number of droplets in a given area and eliminating need for the addition of surfactants or other additives. The low level of water added through fog systems–typically at 0.01 percent to 0.05 percent by weight of the material, or amounts generally less than one pint per ton–generally will not degrade the performance of the material.

There are two methods of producing finely atomized water fog.

Two-Fluid Atomization

One method produces fog from water and compressed air by passing them together through a two-fluid nozzle. Here the external air supply is the vehicle that fractures the water supply into the droplet mist to capture the dust. The supply of compressed air to this system provides an additional expense for the installation and operation of this system. The cost of producing the compressed air must also be considered in the economics of the system. An additional consideration is the possible consequence of injecting additional air into the transfer point's dust control equation, as the velocity of this air can further stimulate the movement of dust.

Single-Fluid Atomization

This system uses an ultra-fine stream of water pumped at high pressure through single-fluid atomizing nozzles. It does not require compressed air or any other additional power source other than the electricity to run its pump. The single-fluid nozzles use hydraulic atomization to generate the fog. Under this method, water is forced under high pressure through a small orifice that shatters the water droplets into microscopic particles. The energy created by the high-pressure pump is used to atomize the water droplets, rather than increase water velocity, thereby minimizing displaced air. By eliminating the compressed air requirement, the single-fluid nozzles simplify installation and reduce operating costs. To keep nozzles clear, any suspended solids must be removed from the water. However, the low volume of water applied to the material makes this relatively easy to accomplish with filtration and electronic treatment.

Location of Fog Systems

The installation of fog systems is a little unusual in that fog systems are designed to treat the air around the material, rather than the material body itself. Therefore, the application point for the fog mist is generally near the end of the transfer point. (**Figure 11.5**) This allows the material load to settle, and any pick-ups for active or passive dust collection systems to remove dust-laden air without risk of blinding the filtration media with moistened dust particles.

Fog generation nozzles are installed to cover the full width of the conveyor's skirted area. (**Figure 11.6**) It is recommended that skirtboard height be at least 24 inches (600 mm) to allow the cone of nozzle output to reach optimum coverage.

Nozzles should be mounted in a way to maximize their ability to fill the enclosure. The nozzle spray pattern should be designed so that all emissions of material are forced to pass through the blanket of fog without putting the fog spray directly onto the main body of material. The spray is directed above the material, rather than at the material.

The spray pattern from the fog nozzles should not directly impinge upon any surface, and the nozzles should be shielded from falling material.

Pluses and Minuses of Fog Systems

Fog systems provide highly effective dust capture combined with economical capital and operating costs. The system operating costs are low when compared to conventional dust collection systems.

A well-designed fogging system can provide excellent control of dust at the point of application without the need for chemical additives. This is especially important for processes such as the transport of wood chips destined for fine paper making. Many mills are very concerned over the application of any chemical that might negatively affect the pulp or degrade the quality of the finished paper.

Since fog systems add only water, they protect the integrity of the process. Total moisture addition to the bulk material can be as low as 0.01 percent to 0.05 percent. This makes fog suppression systems attractive in industries that cannot tolerate excess moisture, such as cement and lime production.

Because of the small orifice size of the nozzles, potable (drinking) water is typically required for fog suppression systems, so filtration to remove suspended solids from the water supply is required. Nozzles can plug if the water supply is contaminated or if the water treatment system is not serviced at required intervals.

If the plant is situated in a cold weather climate, preparations such as drains and heat traced plumbing should be provided.

Another consideration prior to choosing a fogging device is the air volume and velocity at the open area surrounding the transfer point or chute. Fog systems using one-fluid nozzles (without requiring compressed air) tend to be more compatible with the need to control the air movement through transfer points, as discussed elsewhere in this volume. For truly effective performance, fog dust suppression systems require tight enclosure of the transfer point that minimizes turbulent, high-velocity air movement through the system. Since the droplets are very small, both the fog and the dust can be carried out of the treatment area onto surrounding equipment by high-velocity air leaving the chute. Fog suppression works well where the area to be treated is not large.

Disadvantages of fog generation systems include the fact that system design and installation require some degree of experience and some degree of customization –both of which can add to the system's price tag.

Another potential drawback of a fogging application is that this form of dust treatment is site specific. That is, dust control is achieved only at the point of application. Several fogging devices may be required for a complex

Figure 11.6
Fog application nozzles are installed to cover the full width of the conveyor's skirted area.

conveyor system with multiple transfer points. The overall capital expenditure may preclude fogging if the conveyor system is too extensive.

Adding Chemicals to Water

It is a common practice to "enhance" the water by adding chemical surfactants. This addition will improve the wetting characteristics of water, reduce overall water usage and minimize the drawbacks associated with excessive moisture addition.

If dust from coal, petroleum coke, or a similar material falls onto a puddle of water, the dust particles can–if undisturbed–lie on top of the pool for hours. This phenomenon takes place because these materials are hydrophobic; they do not mix well with water. Since it is not possible or practical to alter the nature of the dust particles to give them greater affinity for water, chemicals are added to alter the water particles so they attract or at least join with the dust particles that they contact.

By adding chemicals (usually surfactants; that is, surface acting agents) the surface tension of the water is reduced, allowing the dust particles to become wet.

Surfactants are substances that when added to water, improve the water's ability to wet surfaces and form fine droplets. Surfactants lower the water's surface tension and overcome the internal attraction between the molecules of water, ultimately resulting in an improved droplet formation.

To understand surface tension, imagine a drop of water lying on a smooth, flat surface. It will usually form a liquid bubble with well-defined sides. It is the surface tension of the water that prevents the droplet walls from collapsing.

A drop of water that has been mixed with a surfactant such as dishwashing soap, for example, will not form a liquid bubble on the same surface because its surface tension has been drastically reduced. The "walls" on the side of the droplet cannot support the weight of the droplet, because the forces holding the walls together have been altered. This is the reason surfactant technology is applied to dust control. If the water droplets no longer have a surface that is a barrier to contact with the dust fines, then random collisions between droplets of treated water and dust will result in wetting of the fines.

Choosing a Surfactant

The number of surfactants and surfactant blends currently in use is quite extensive. A number of specialty chemical companies have products formulated to address specific dust control needs. Choosing the correct product and addition rate for a given application requires material testing as well as an understanding of the process and the application.

Objections to chemical additive-enhanced water suppression systems include the continuing cost to purchase the chemical additive, at prices typically ranging from one-half cent to three cents per ton of conveyed material. These costs can be higher, particularly when considering the amortization of equipment. In addition, these systems require regular maintenance, which adds labor expenses to the consideration of continuing costs.

As contamination of the material or process can be a great concern in some industries, the additive chemical must be reviewed in this light. It is important that the chemical additives are compatible with the process, the bulk materials, and with system equipment, including the conveyor belting. Although the use of a surfactant reduces the amount of water added to the dusting material, water/surfactant sprays may still add more water than is acceptable.

Application by Foam or Spray

Once the most efficient wetting agent has been selected, the decision must be made whether to apply the material as a wet spray, as discussed above, or as foam. Both systems offer advantages. Generally speaking, the moisture addition from a wet spray system is higher than that of a foam-generating system. Although the dilution fate is lower for the foam suppression, **(Figure 11.7)** the expansion of the foam allows it to provide effective suppression with lower moisture added to the material. Recent developments in surfactant technology have improved to the point some blends can be applied at the low-moisture levels of a foam system while providing good dust suppression. This would provide the benefit of the limited moisture addition while minimizing chemical cost due to the higher dilution rates with the spray-applied surfactants.

Foam Dust Suppression

The use of surfactants with water will improve the likelihood that fines will collide with droplets, and the collisions will result in suppression of the dust. It stands to reason that the next objective would be to maximize the surface area of the available water droplets to make as much contact with the dust fines as possible, thus limiting the amount of water added.

Some suppliers offer dust suppression systems that create a chemical foam. Given the proper mixing of material and foam, these systems provide the most effective dust control. As the moisture is in the form of foam, its surface area is greatly increased, improving the chance for contact between dust and water. Some foam bubbles attract and hold dust particles together through agglomeration. Other bubbles implode on contact with dust particles, releasing fine droplets that attach to the smaller, more difficult to catch, and more hazardous dust particles.

With moisture addition of 0.05 percent to 0.1 percent, foam suppression systems typically add less than 10 percent of the water used by a water-only spray suppression system. These figures translate to approximately one quart of water per ton (a little more than one liter per metric ton) of material. Consequently, these systems are welcomed where water supplies are limited, or where excess water can downgrade material performance, as in power plants. In addition, the reduced water means fewer problems with screen clogging and adherence of material to mechanical components and walls.

Some surfactants lend themselves quite well to creating foam. Adding air to the surfactant and passing

Dust Suppressant Chemical Dilution Rates	
Chemical Type	Dilution Rate
Foam Surfactant	100:1 to 250:1 ((1% to 0.25%)
Wet Spray Surfactant	700:1 to 1,500:1 (0.14% to 0.07%)

Figure 11.7

While foam surfactants are diluted less, the expansion of the foam adds less moisture to the material.

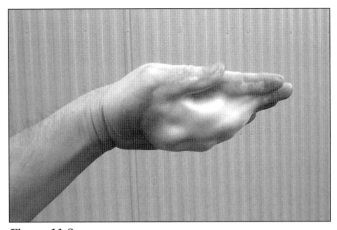

Figure 11.8
Foam dust suppression generates a "dry" foam that expands the surface area of water from 60 to 80 times.

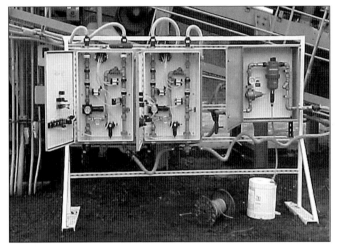

Figure 11.9
In the proportioning system, water and surfactant are mixed and the resulting solution and compressed air are sent independently to the foam canisters.

Figure 11.10
The water/surfactant solution and air are combined in the foaming canister and supplied to the application nozzles.

this compound through a mixing device creates foam. Adjustment of the air/water/chemical ratio and other controllable factors allow the application engineer to generate foam ranging from very wet to "shaving cream" dry, depending on the application needs. A well-established foam can expand the surface area of a given quantity of water 60 to 80 times. This allows for much lower rates of moisture addition. **(Figure 11.8)**

Water quality plays an important role in the effectiveness of a foam dust control program. The ability to generate acceptable foam is largely dependent on the quality of the water used. In order to ensure that foam generation can take place at a given site, it is prudent to analyze the plant water for the following characteristics:

- pH
- Conductivity
- Total Suspended Solids
- Calcium Hardness

Each suppressant supplier has a specification for these attributes. The vendor will also determine suitable foaming agents and dilution rates based on these characteristics.

Foam Generation and Application System

The system for the application of foam dust suppression begins with the mixing of the water and foam-generating chemical. Water and additive are metered together through a proportioning pump, and the resulting mixture is pumped through a flow regulator to feed the system. **(Figure 11.9)** A second flow regulator controls the flow of compressed air. The solution and air arrive via separate hoses at the foaming canister, where they mix together. The foam created travels through hoses to the application nozzles installed in the wall or ceiling of the chute at the application point. **(Figure 11.10)** One proportioning pump-and-regulator system can feed two points of application in plant.

Limitations of Foam Suppression

While many applications can benefit from foam technology, there are some liabilities to the process. In many cases, the surfactants that produce the most desirable foam qualities are not the best wetting agents for the material to be treated. Some suppliers focus on chemicals to produce stable foam without considering whether the resultant foam is of any value in overcoming the hydrophobic nature of the fines. It is critical that the chemicals provide effective wetting performance before foam generation is considered.

Foam generation requires compressed air. If compressed air is not readily available at the application site, a compressor must be installed and maintained. The foam application equipment is slightly more expensive than conventional spray equipment and requires somewhat more maintenance.

Finally, the amount of surfactant required to generate foam is somewhat greater than the amount required for wet spray. (The volume of surfactant to a given body of water is higher; however due to the foam's expansion, the amount of mixture applied is lower (as shown in **Figure 11.1**.)

The cost of the additive chemical may be offset by a reduction in BTU penalty on fuel performance resulting from a substantial decrease in added moisture.

Residual Chemical Suppression Agents

Another dust suppression technology incorporating chemicals uses binders and residual agents to extend the dust suppression effect for longer periods.

The use of non-residual surfactants allows for the wetting of dust fines so that the fines agglomerate together or attach themselves to larger particles, thereby, preventing them from becoming airborne. Once the water used to wet the fines evaporates, the dust suppression effect is gone.

In many cases, however, dust suppression is required through multiple transfer points until the material reaches storage bins, railcars, barges, or production sites where the need for dust control ends. For instance, coal that is conveyed from rail to barge to open storage piles can remain in storage for several weeks. When this coal is reclaimed, it may be very dry and present a greater dusting problem than it did during its initial handling. In other cases, very dusty materials such as, calcined coke or iron ore pellets, may require dust control from the point of production to the point of end use, which may be several weeks and several thousand miles apart.

In such cases, it is sometimes more economical to apply a residual surfactant/binder to the dusting material rather than to apply surfactants and water at multiple (repeat) application sites throughout the application process. There are a variety of residual binders available to suit almost any need.

The objective of a residual binder is to agglomerate fines to each other or to larger particles and then hold the structure together after the moisture evaporates. In some cases, a hygroscopic material, such as calcium chloride, is used, which retards the ability of moisture to leave the treated material. The advantage to this approach can be a low treatment cost.

More conventional binders include lignin, tannin, pitch, polymers, and resins. When combined with surfactants to aid wetting, these compounds coat the larger particles that then act as a glue to attract and hold dust fines. The success of these programs depends on the dust suppression application and mixing of the treated material to ensure that the dust fines have come into contact with the larger treated particles.

Application of residual binders tends to be more expensive than surfactant applications, because they must be applied at a lower dilution. Although binders are less expensive per pound, they are typically applied at dilution rates ranging from 50-to-1 to 200-to-1 (2 percent to 0.5 percent).

When choosing a binder, it is especially important to know the effects the binder will have on transfer equipment and conveyor belts. If the binder adheres well to the dusting material, it may do the same to the handling equipment. Proper application of the product becomes critical, because overspray of the binder onto process equipment or empty belts can result in considerable production and maintenance problems.

An important consideration in selecting a binder is the effect of this chemical on both the environment and the material being treated. When a binder is applied to a material that goes into storage, the storage pile may be exposed to rain. If the binder is water soluble, a portion of the binder may go with the runoff and provide an environmental concern. Most chemical manufacturers provide only those binders that are environmentally safe; however, this is an issue that should be raised with the supplier.

Location, Location, Location

In fog, foam, water, and water/chemical spray applications, the sites chosen for nozzle placement and suppressant delivery patterns are as important, if not more, than the selection of the material to be applied. **(Figure 11.11)** Even the best-designed program will fail if the suppressant material is not delivered to the correct location to allow intimate mixing of the suppressant with the dust fines.

The success of the suppression effort relies on the proper mixing together of the material and the suppressant at the transfer point. When applying dust control, whether the suppressant is simply water or a surfactant/water mixture, as a spray or a foam, it is best to locate the suppression system as close to the beginning of the transfer point (and dusting sources) as possible. That way, the forces of the moving material fold the suppressant into the material body as it moves through the transfer point. For example, foam suppression is usually most effective when applied at the discharge point of a crusher or conveyor, where the body of material is in turmoil and expanding. **(Figure 11.12)** The application of the suppressant at this point allows the foam to penetrate more deeply into the body of material and capture individual particles, rather than being held on the material's external layer.

Controls

The application of dust suppression water and/or chemicals at transfer points must be controlled

Figure 11.11
Foam suppression is most effective when applied where the material is in turmoil, as at the discharge of a crusher or conveyor.

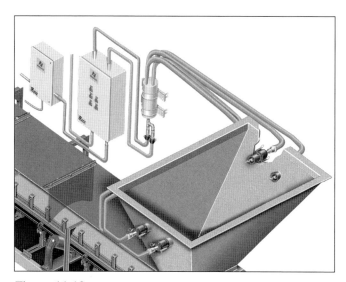

Figure 11.12
The location of the application point is critical to the success of any dust suppression system.

automatically, so water and/or chemical is applied only when the conveyors are running, and there is material present.

Control systems can be as simple as an operator opening a valve on a "garden hose" system. At the other end of the spectrum can be a stand-alone computer system dedicated solely to monitoring and controlling the dust control system.

As the control systems become more sophisticated, the operating cost of the dust control system goes down. More sophisticated control systems allow greater latitude in the selection and operation of sensors that can detect material motion, material throughput, instantaneous dust loading in the area, chemical usage, air and water flows, and even the percent of moisture being added to the material.

Sensors that can differentiate between two different materials can allow the suppressant type or flow rate to be changed automatically as the material changes.

Dust Testing

In the United States, the Environmental Protection Agency (EPA), Mine Safety and Health Administration (MSHA), Occupational Safety and Health Administration (OSHA), and other governmental agencies at federal, state, and local levels are now requiring dust level testing to ensure compliance with current laws.

This testing is typically done over a period of time, and the results weighted to reflect the duration of the test. In many instances, it is better to use testing equipment that will monitor and record dust levels on an instant-by-instant basis. This will provide a more accurate indication of dust problems and how they relate to material movement. For instance, many crushers have very low dusting levels when choke-fed, but dust levels rise to unacceptable levels when the material flow is slowed or cut off. Using the time-weighted testing method will show total dust output, rather than the high dust levels produced during the short intervals.

By using the instantaneous type of dust monitor, these short-interval, high-level dust load areas can be pinpointed, and actions can be taken at these specific problem areas to reduce dusting problems.

System Maintenance

Just as with operating an automobile, dust suppression systems require routine maintenance. Without a doubt, the most common cause of failures for dust suppression systems is a lack of service.

Nozzles must be checked and cleaned, filters cleaned, chemical levels checked, and flow rates for water and air checked on a routine basis, or even the best system is doomed to failure.

Some suppliers of dust suppression equipment and chemicals are now offering this routine service as a part of their system package. It is wise to consider this solution, as it will free maintenance and operating personnel for other duties and guarantee the integrity of the dust suppression system.

Dust Suppression: One Piece of the Puzzle

There is no such thing as 100 percent dust control. Even "clean rooms" in laboratories have some dust in them. Be realistic in expectations for dust control and look for specific percentages of reduction instead of adopting a "no dust allowed" approach.

Dust suppression is best suited to enclosed spaces of reasonable size. It becomes difficult to apply and control any of the various forms of dust suppression in open areas or inside large structures, such as railcar dumps or truck dump sheds as seen in mining applications. A combination of suppression and collection can be applied in these large situations to achieve acceptable results.

Dust suppression alone cannot be the complete answer to the control of fugitive material. But it carefully chosen, effectively engineered, and properly maintained, a dust suppression system can improve the efficiency and minimize the risk of overloading its own components, the material containment, and dust collection systems. Effective dust suppression is a critical portion of the pyramid of total material management.

chapter 12

Dust Collection

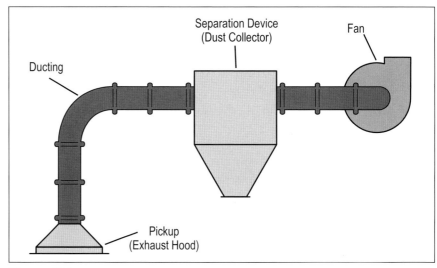

Figure 12.1
Basic Components of a Dust Collection System.

"Mechanical dust collection systems are installed to pull dust-bearing air away from the conveyor transfer point, separate the dust, and then exhaust the air."

Dust collection–the passing of dust-carrying air through some form of filtration or separation system–presents the final piece in our dust control system. There are both active and passive dust collection systems. A passive system merely allows the air to move through the filtration system, while active systems like a vacuum cleaner to pull in the air through a filtration medium to remove the solids.

Among the considerations in the selection or engineering of a dust collector system are:
- Concentration (amount) of dust and the particle sizes.
- Characteristics of the dust (abrasive, sticky, hygroscopic, combustible, etc.).
- Characteristics of the air stream (temperature, amount of water vapor, and/or chemicals present).
- Degree of collection efficiency required.
- Final disposition of the dust (how it is to be removed from the collector, and how it is to be disposed of or reused).

Active Dust Collection Systems

Mechanical dust collection systems are installed to pull dust-bearing air away from the conveyor transfer point, separate the dust, and then exhaust the air. A typical dust collection system **(Figure 12.1)** consists of four major components:

- Exhaust hoods (or pickups) to capture dust at the source(s).
- Ductwork to transport the captured air/dust mixture to a collector.
- A collector, filter, or separation device to remove the dust from the air.
- A fan and motor to provide the necessary exhaust volume and energy.

There are three basic approaches to dust collection systems: central, unit, or integrated systems. Each has its own advantages, depending on the specifics of the application.

Central Method

The central method of handling the total air for a conveyor system would be to connect all the individual collection points by means of ducting to a single dust collector that is installed at a single, more-or-less remote location. (**Figure 12.2**) This collector contains fans, filters, and a collection hopper. This filtration system would handle all the dust extracted from the entire conveying system, collecting it for disposal, or feeding it back onto the conveyor or into the process at a convenient point.

Central systems are particularly suitable when the process being exhausted has all dust-generating points operating at one time, and/or it is desirable to process all dust at one site. It is also useful when there is limited space for dust collection and processing equipment installation near the conveyor or where the explosion risk requires the dust collector to be positioned at a safe distance. The drawbacks of the central dust collection system are its requirements for more complex engineering and the need for lengthy systems of ducting. As all dust gathering points must be operating at once, the central method may present higher operating costs. In addition, to service any one component, the entire system must be shut down.

Unit Method

This concept consists of small, self-contained dust collectors for individual or small and conveniently grouped dust generation points. (**Figure 12.3**) The dust collector units are located close to the process machinery they serve, reducing the need for ducting. Typically, the dust collection units employ fabric filter collectors for fine dust; cyclone collectors are normally used with coarse dust.

The unit system provides the benefit of reduced ducting and the resulting reduction of engineering and installation work, as well as reduced operating costs since portions of the operation may need to run only intermittently. Each unit can be serviced independently

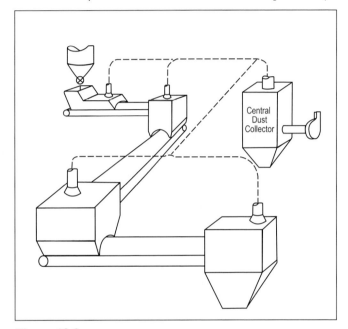

Figure 12.2

A central system uses a single collector to remove dust from a number of different points or operations in the plant.

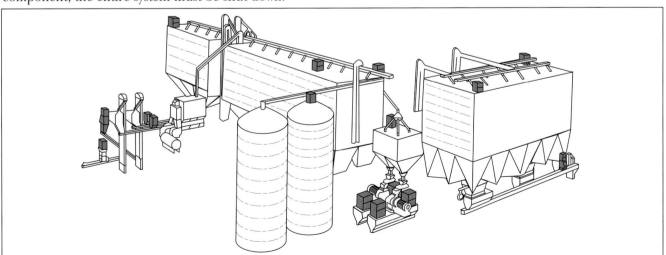

Figure 12.3

The unit system places smaller dust collectors near individual or closely grouped dust generation points.

without needing to shut down the whole system. The unit method does require more space adjacent to the dust sources. The disposal of dust from each of the unit collectors can require additional dust handling mechanisms.

Integrated Systems

A logical extension of the unit concept is the integrated system, where the dust control system is incorporated within the dust generation point itself. **(Figure 12.4)** The integrated approach has been made possible by the development of the insertable filter. This filter is built into the enclosure around the dust-creation point, so the dust is controlled at its source. The dust is not "extracted" at all; it is collected and periodically discharged back into the material stream.

A principle advantage of this system is a reduction in the need for ducting. The integrated system is often more economical than centralized or unit systems, unless there are many points in proximity that require dust control. It operates only when needed, which reduces energy requirements. Because it can return the dust right to the process, there is no need for a separate dust disposal system. If the method of returning the dust to the source uses compressed air, it can cause a puff of dust at the material entry and exit areas.

Dust Separation Technologies

There are a number of specific approaches to removing the dust from the air, each with its own benefits and drawbacks.

Baghouses

Perhaps the most common dust separation technology is the use of fabric collectors, which are placed in structures commonly called baghouses. **(Figure 12.5)** In this system, dust-laden air enters an enclosure and passes through bags of filter fabric. The bags–of woven or felted cotton, synthetic, or glass-fiber material in a tube or envelope shape–need to be cleaned periodically by mechanical shaking, reverse air, or reverse jet.

There are three basic rules of baghouse operation:
1. Optimum cleaning efficiency depends on the dust cake buildup on the filter surface. Performance is better on a filter with some cake buildup, rather than a brand new filter.
2. Airflow quantity depends on the filter medium's permeability, the amount of buildup before filter cleaning, and the power of the blow.
3. The more permeable the cloth, the less efficient its collection, with or without a dust cake.

With the mechanical shaking design, the air passes

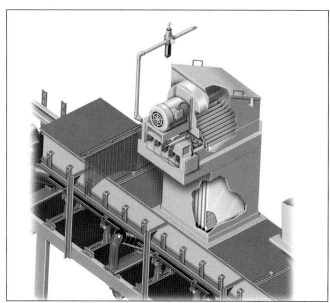

Figure 12.4
Integrated systems place insertable dust collectors within the dust generating equipment such as conveyor transfer points.

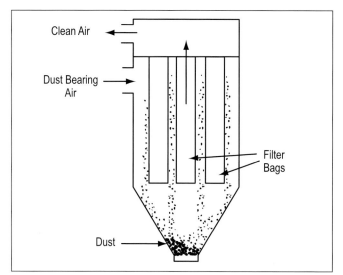

Figure 12.5
"Baghouse" Dust Collector.

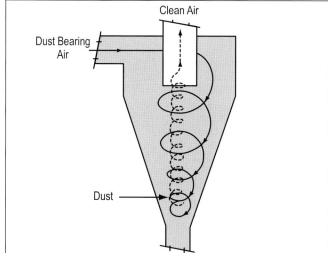

Figure 12.6
Cyclone Dust Collector.

from the inside to the outside of the bag, with the dust captured on the inside of the bag. The bags are cleaned by shaking the top mounting bar from which the bag is suspended. The system needs to be stopped for cleaning.

In the reverse air style, the bags are fastened at the bottom. Air moves up through the bag from inside, with the material collecting on the inside. Bags are cleaned by injecting clean air into the dust collector in a reverse direction, so the bag partially collapses, causing the dust cake to crack off the bag wall and fall into the hopper bottom. Again, the system needs to be stopped for cleaning.

In reverse jet systems, the bags are fastened from the top of the bag house and supported by metal cages. Dirty air flows from outside to inside the bags, leaving the dust on the outside of the bag. Bags are cleaned by discharging a burst of compressed air into the bags, which flexes the bag wall and breaks the dust cake off so it falls into the collection hopper. These systems can run continuously while being cleaned.

The bag house collection system is up to 98 percent effective in removing respirable dust emissions. The filtration bags are relatively inexpensive compared to other methods, and the large number of manufacturers ensures a competitive pricing in the marketplace. The disadvantages of these systems include problems in high-temperature applications and high-humidity conditions. Some systems require entry into the bag house for replacement of the filter bags, with employee exposure to high dust levels and explosion problems as major concerns.

Cartridge Filter Collectors

Instead of bags, this system places perforated metal cartridges containing a pleated, non-woven filter media inside the structure. These systems are available in single-use (change the filter while off line) systems and with pulse-jet cleaning systems, allowing continuous-duty cleaning.

The pleated filter media used in this system provides a larger collection surface in a smaller unit size than other dust collection systems. As a result, the size of the overall system can be reduced. However, replacement cartridges tend to be expensive. High moisture content in the collected material may cause the filter media to blind, and the system itself requires fairly high levels of maintenance.

Cyclones

These systems create a vortex–an internal tornado–that "flings" the dust out of the air stream. (**Figure 12.6**) The whirling airflow created inside the structure creates centrifugal force, which throws the dust particles toward the unit's walls. After striking the walls, particles agglomerate into large particles and fall out of the air stream into a collection point or discharge outlet at the bottom of the unit.

There are single cyclone separators that create two vortexes in one enclosure. The main vortex spirals coarse dust downward, while a smaller inner vortex spirals up, carrying fine particles upward toward a filter. Multicyclone units contain a number of small-diameter cyclones operating in parallel, with a common inlet and outlet. These units create the same two vortexes in each cyclone chamber. They tend to be more efficient, because they are taller, providing greater time for air inside, and smaller in diameter, providing greater centrifugal force.

Due to their lack of moving parts, these systems are low maintenance. They are often used as pre-cleaners to reduce the workload on more efficient dust collection systems. They do not provide efficient collection of fine or respirable particles, and performance suffers in high-

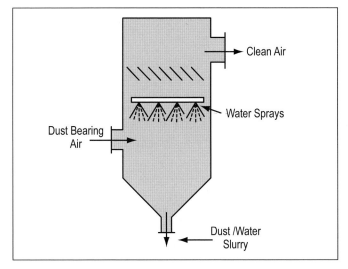

Figure 12.7
Wet Scrubber.

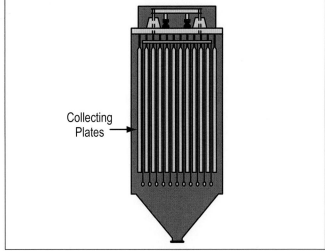

Figure 12.8
Electrostatic Precipitator.

humidity conditions. They need to maintain a high rate of air flow to maintain the separation process.

To improve the efficiency of some cyclones–particularly with fine particles–the collecting surface of these units may be wetted with water.

Wet Scrubbers

In these systems, a liquid–most commonly water–is sprayed (usually down) into the stream of dust-bearing air. **(Figure 12.7)** The dust particles are captured by water droplets and fall out of suspension in the air. The dust-and-water mixture is released out the bottom of the collector as a slurry, which is then passed through a settling or clarification system.

These systems are relatively low in installed cost and can be used in high-temperature applications. On the downside, these systems do have higher operating and maintenance costs and may require freeze protection for cold weather operations. They may need a pre-cleaner (such as a cyclone) for heavy dust conditions, and water treatment is usually required for the contaminated water discarded by the system.

Electrostatic Precipitators

These systems apply a negative electrical charge to the particles as they pass into the collectors. **(Figure 12.8)** The charged particles are then attracted and adhere to positively-charged electrode plates. Rapping or vibrating the electrodes then discharges the dust.

As the need to attract small particles grows, the charge must become larger, and the ESP chamber must also get bigger.

Precipitators can be 99 percent effective on all dust, including sub-micron particles. They will work on all materials, including sticky and corrosive material, in high-temperature or in high gas-flow environments with minimal energy consumption. However, they require a large capital investment to obtain. In addition, safety measures are required to prevent exposure to the system's high voltage. They can allow or cause an explosion when combustible gases are collected around the electric system.

Filter Material and Size

After specifying a style of dust collection system, the next step is selecting the material for the filter medium, if required.

Selecting a filter of the correct material and size is a critical function. Technology has brought us many new and advanced filter media. These advancements allow the designer to pinpoint the proper style and material for the filter. Again, selection should be based on the type of dust to be collected. For example, if the collected dust is at a temperature that is borderline of a standard filter, a high-temperature medium can be selected. This same material can also be chemically treated to resist moisture.

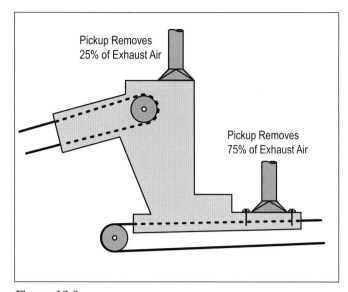

Figure 12.9

Conveyor transfer points typically require several dust pick up points.

Cleaning efficiencies of filter systems as well as medium design are critical to the success of a dust collection system.

Many filter manufacturers publish lists of filtration velocity–that is, the volume of air in cubic feet divided by the effective area of filtration in square feet–for the various materials. However, these lists should serve not as absolute rules, but rather as guidelines, to be modified by variables such as dust particle size, process temperature, and moisture presence.

It is worthy of note that most filter media need a dust cake on the surface to attain the desired collection efficiency. All too many people think that the cleaner the bags are, the lower the emissions. This is not true; it is possible to over clean the dust bags. Dust collector manufacturers typically offer a valuable option in a device that pulses the baghouse only when the pressure drops.

Dust Pickup Locations

As the old saying goes, "the three most important things in retail business are location, location, and location." The same is true in dust collection: a critical element is the location of the pickup points.

It is important that the material fines be allowed a chance to settle, either on their own accord or with the addition of dust suppression–in the form of a spray or fog system–before dust collection points are reached. Otherwise, energy will be wasted removing dust that would have shortly settled on its own, and the dust suppression and collection systems will be larger (and hence more expensive) than would be required. The location should prevent capturing coarse particles, which settle quickly, and instead capture only fine, predominantly respirable dust.

To prevent dust from escaping from the container's entry, sufficient air must be extracted by the dust collection system to match the air influx in the stream of material (induced and generated air) and the air displaced by the entering particles (displaced air).

Multiple dust collection pickup points are usually required. Often, one dust collection pickup is located at the entry area directly before the load zone to take in approximately one-fourth the total calculated air movement. The second dust collection pickup point is positioned approximately two times the belt width after the load point to collect the remaining three-fourths of the air movement. **(Figure 12.9)**

In their reference text *Industrial Ventilation* the American Conference of Governmental Industrial Hygienists has prepared standard calculations for conveyor belt ventilation. They specify that belts operating at over 200 feet per minute (1 m/sec) require an air evacuation rate of 500 cubic feet per minute per foot of belt width. (This translates into metric measurements as 7.8 liters per second per centimeter of belt width.) The exhaust port should be installed above the enclosure of the belt receiving the load. This exhaust port should take all or most of the exhaust air from the receiving belt enclosure. If the height is over 10 feet (3 meters) or if a rock box is installed in the chute, the system may require removing air from two ports, taking two-thirds of the air at the receiving belt and one-third of the air at the discharge belt.

If the material drops more than three feet (1 meter) onto the belt, *Industrial Ventilation* calls for an additional exhaust system, again removing 500 cubic feet per minute per foot (7.8 liters per second per centimeter) of belt width. This secondary exhaust should be placed approximately one-third the belt width from the end of the sealed skirtboard section.

Industrial Ventilation also notes that dry, very dusty material may require exhaust volumes 1.5 to 2.0 times these stated values. These dry materials may require an additional exhaust of 700 cfm (330 l/second) for belts from 12 to 36 inches (300 to 900 mm) or 1000 cfm (470 l/second) for belts above 36 inches (900 mm), installed on the back or pulley side of the loading chute.

The above calculations are general estimates and, as such, should be reviewed by the designer of dust collection systems for specific applications.

The Risks of Fire and Explosion

As evidenced by silo explosions in the grain handling industry, dust explosions are very powerful and a very real risk. Consequently, extreme care must be taken to minimize this risk.

For many dusts, a settled layer as thin as 1/32 of an inch (0.75 mm)–the thickness of a paper clip–is enough to create an explosion hazard. A 1/4-inch (6 mm) layer is a bigger problem: big enough to destroy a plant.

For there to be a dust explosion, three factors need to be present: a combustible dust at the right concentration, a gas that supports ignition, and an ignition source.

Many fine dusts, including chemicals, food products, fertilizers, plastics, carbon materials, and certain metals, are highly combustible. By nature, any dust collection device contains clouds of these fine particles suspended in air, which itself is a gas that supports ignition.

In any mechanical material handling operation, there are a number of possible ignition sources:

- *Mechanical Failures* that cause metal-to-metal sparks or friction.
- *Overheating* from a worn bearing or slipping belt.
- *Open Flames* from direct-fired heaters, incinerators, furnaces, or other sources.
- *Welding or Cutting* causing a point source ignition or a hot particle dropping (perhaps several floors) to a flammable atmosphere.
- *Static Electricity*.
- *Others* including flammable dust entering the hot region of a compressor or catalytic reactor.

Categorizing Dust Explosions

There are several ways to look at dust-related conflagrations.

Flash Fire
> This is the ignition of unconfined dust. This free dust will not destroy a building with explosive force, but can ignite other combustibles and cause significant damage to structures and equipment.

Explosion
> When dust is confined and ignited, an explosion is created. This rapid explosion of gases will generate significant and destructive over-pressures that can even demolish the building, leading to greater damage and injury.

Primary or Secondary
> An initial or primary explosion can cause secondary explosions by disrupting, dispersing, and igniting new sources of dust removed some distance from the original blast. Secondary explosions can be more destructive than the primary explosion, and every explosion can lead to additional secondary explosions.

Magnitude
> The speed and force of an explosion are direct functions of a measurable characteristic called the deflagration index. Dust explosions can be more hazardous than explosions caused by flammable gases.

Control Mechanisms

In the presence of these ingredients, precautions must be taken to avoid a dust explosion. Control mechanisms include:

Inerting
> the addition of an inert gas (typically nitrogen or carbon dioxide, rather than air) into the collector.

Suppression
> adding a suppressant material as explosive pressure starts to rise.

Venting
> adding an explosion relief-panel or bursting membrane, which releases the explosion energy out of the enclosure.

Proper grounding of the filter will help reduce the risk by increasing conductivity through the system, allowing static charges to leak into the ground.

Consult with equipment suppliers to design dust collection systems for handling potentially explosive dusts.

Venting

The theory behind explosion vents is simple. It is a deliberately weakened wall that will release early in the pressure rise created by a rapidly rising temperature. Once this weakened area is opened, the burned and unburned dust and flame can escape the confined area, so the vessel itself does not experience the full rise in pressure. If the release is early and large enough, the pressure will remain low inside the vessel to protect it from damage. However, the fire or explosion can develop outside the vessel and if dust is present, other equipment can be damaged, so venting systems do not eliminate the need for good dust control systems and standards of housekeeping.

Figure 12.10
This conveyor transfer point features an insertable dust collector.

There are two types of explosion-venting devices. Rupture disks are thin panels that will open faster than other designs. They must be sized to withstand normal negative operating pressure (typically 8 to 12 inches of water), yet rupture at a positive explosive pressure. A more widely used design is the spring-set door. This door–available either in hinged or unhinged designs–will vent (pop open) during a conflagration.

Handling Of Collected Material

The final requirement in any dust control system is to provide a mechanism to dispose of the dust after it has been collected. Operations that must be considered include removing the dust from its collector, conveying the dust, storing the dust, and treating the dust for re-use or disposal.

The handling of collected material can be a problem, particularly if the material is to be returned to the process. Care must be taken to avoid affecting the process through the introduction of an overload of fine particles at any one time. In addition, it is important the collected dust be returned into the main material body in a manner that avoids the re-energizing of the dust so it is collected again at the next pickup point. Because the dust particles are small enough to readily become airborne again, they are often subjected to an extra combining process. Collected dust can be put through a mixer, pug mill, or pelletizer before it is re-introduced to the general material handling system. In some industries, however, collected dust is considered contaminated and, therefore, must be sent to a landfill or otherwise properly disposed of as waste material.

Downfalls of Dust Collection Systems

Dry collection systems to clean dust-laden air work well in both warm and cold climates. But these dust systems–regardless of the selection of central, unit, or integrated approach–require a large amount of space for equipment and ductwork, making them expensive to install. And of course, the operating and maintenance costs are multiplied as the size of the system increases. Changes or alterations required after start up may be hard to implement without modification of the entire system. Even filter bag replacement can be costly and time-consuming. The collected dust must be returned to the material flow (or otherwise disposed of), which may allow the dust to be re-entrained into the air and then re-collected at the next pickup point.

The Importance of Planning for Dust Control

Dust suppression and collection systems should be considered early in the design stage of the conveyor and in the planning of the entire operation. Best results are generally not obtained if dust control is an afterthought.

Correctly applied dust collection equipment can boost the efficiency of the entire operation as well as provide a high degree of environmental protection. **(Figure 12.10)** Effective dust control will avoid unnecessary waste of the material, protect sensitive equipment, and minimize any potential explosion risk, while ensuring compliance with environmental and workplace regulatory requirements.

Equipment that is of good quality and adequately sized should always be used. The temptation to apply the smallest, cheapest equipment should be resisted, as inadequate dust control systems will turn into an ongoing liability, rather than an asset. It would be desirable if the installed system was not fully taxed into operating at the top end of its performance levels at installation. It would be better to pay a little more now for a modestly oversized system. This "growth room" allows for changes in material volume or characteristics, more stringent regulatory requirements, or other operational changes.

This section has presented only an overview of the capabilities and considerations of dust collection and control equipment. It would be wise to consult with suppliers specializing in this equipment to receive specific recommendations.

An effective dust collection system need not be expensive. In fact, it always adds to the company's bottom line. It protects the owner, the employee, and the environment, as well as satisfies the insurance carrier, the fire department, and regulatory agencies, such as OSHA in the plant and EPA off premises. Equally important, effective dust collection increases product quality, enhances finishing operations, helps machinery and computers operate trouble-free, and if designed correctly, is the cheapest method available for handling significant amounts of a waste so common in industry.

The Systems Approach To Dust Management

It is obvious that the more successful the enclosure and sealing, the less fugitive dust created. The more successful the dust suppression system, the less dust in the air to be captured. Successful application of the principles of enclosure and suppression works to minimize the wear and tear and reduces the risk of overload on the dust collection system. A pyramid composed of these three systems–enclosure, suppression, and collection–allows plant management to be successful in controlling the amount of dust released into the environment.

Chapter 13

Belt Cleaning

"Carryback–that is, the material that adheres to the belt past the discharge point at the head pulley–accounts for a sizable portion of the fugitive material present in any plant."

Figure 13.1
Ineffective belt cleaning will allow significant quantities of carryback to accumulate under the conveyor.

Generally speaking, the normal process of carrying material on a conveyor run results in separation of the load into a layer of wet fines resting on the belt; with the coarser, dryer material above the fines; and the largest lumps above that. The lumps tend to discharge in the normal trajectory, while the layers of coarse grit and fines cling to the belt. This residual material is carried back on the belt's return run. As they dry or are dislodged by the vibration of the return rollers, the particles will eventually fall off the belt, accumulating in piles under idlers and pulleys. **(Figure 13.1)**

Carryback–that is, the material that adheres to the belt past the discharge point at the head pulley–accounts for a sizable portion of the fugitive material present in any plant. And like transfer point spillage, carryback can present serious problems for conveyor systems. These problems create consequences in maintenance, downtime, and plant efficiency. They present themselves as accumulations on rolling components, leading to belt mistracking, premature idler failures, and other problems that create conveyor downtime and require unexpected, premature service. Considerable man-hours and expense will be required in a never-ending cleanup of fugitive material originating as carryback. In addition, accumulations of fugitive material on the ground or clouds of

Figure 13.2

Carryback dislodged by the return idlers will form piles along the conveyor.

Figure 13.3

The accumulation of material on return idlers and the conveyor structure poses the threat of belt damage.

dust in the air can present a fire or explosion hazard; a slip, trip, and fall hazard, a health hazard; or a lightning rod for unwanted attention from neighbors or regulatory agencies. No matter where it lands, carryback is a leading cause (and indicator) of conveyor inefficiencies.

To prevent–or more realistically, control–the damage and expense carryback can create, conveyor belt cleaning systems are installed. Typically, these are some form of scraper or wiper device mounted at or near the discharge pulley to remove the residual fines that adhere to the belt as it passes around the head pulley.

The Rise of Engineered Cleaning Systems

For many years, belt cleaners were homemade affairs–often a slab of rubber, used belting, or even lumber wedged against the structure or held against the belt by a counterweight. These systems proved to be unwieldy, cumbersome, and generally ineffective, at least over the long term. Over time, plant operating requirements necessitated the use of wider, faster, more deeply troughed, more heavily-loaded conveyors. These led, in turn, to the development of engineered belt cleaning systems to protect the plant's investment and extend the service life of the more expensive belts and other conveyor components. These cleaning systems were designed to reduce space requirements by allowing cleaner installation in the discharge chute to simplify blade replacement to reduce maintenance time and labor, and to incorporate advanced materials such as plastics, ceramics, and tungsten carbide in order to provide longer service life and reduce maintenance requirements. Also improved were the tensioning devices used to hold the cleaning edge against the belt. Through the use of these engineered cleaning systems, the adhesive mass of fines and moisture traveling at belt speed can be removed from the conveyor.

The Cost of Carryback

In some ways, carryback is a more cumbersome and costly problem than transfer point spillage. Spillage creates a local problem. The fugitive material is seen in the environs of the transfer point, and so the mess it creates can be removed at a relatively low cost.

Carryback, on the other hand, presents a linear problem. **(Figure 13.2)** Fugitive material is released from the belt by the abrasion from the return idlers all along the return run of the conveyor. This makes it a larger (or at least longer geographically) problem, and makes its removal more extensive and expensive than a localized spillage cleanup. Cleaning crews and equipment will need to work along the full length of the conveyor, rather than stay in one area.

In addition to the time and expense of cleanup, carryback creates other problems through increasing the wear and shortening the service life of other conveyor components, including pulleys and idlers. Released carryback material will build up on idler bearings and underneath the rollers, increasing friction and leading to a bearing failure. **(Figure 13.3)** A large German lignite-mining firm calculated that it replaced 12 percent of the return idlers in its operation every year. Roughly 30 percent of those replacements were caused by wear in the support ring or face of the idler, wear that originated in the release of material along the belt line.

If the material is allowed to accumulate, the belt itself will eventually run through the piles, abrading the cover and reducing its service life.

Because carryback accumulates unevenly on the rolling components of the conveyor, it can cause mistracking, which can lead to spillage and off-center feeding onto any receiving conveyor. And of course this mistracking leads to all the other problems that belt wander causes, including shortened belt life, excessive

Figure 13.4

The Stahura Carryback Gauge is a pan with scrapers that is held under the belt to collect a sample of material.

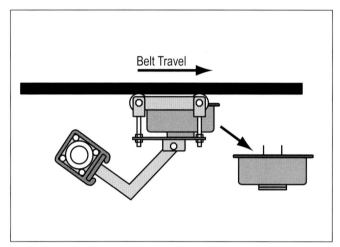

Figure 13.5

Mounting the collecting pan on an arm held in a track improved operator safety.

Figure 13.6

The collecting pan of the ICT Carryback Gauge is moved back and forth under the belt.

labor costs for maintenance and cleanup, unscheduled downtime, and safety hazards.

Measuring Levels of Carryback

The material that sticks to the belt is typically the belt's cargo in its worst state. Carryback particles are finer and have a higher moisture content than is characteristic of the material in general. Due to vibration of the belt as it rolls across the idlers, fines tend to sift down and cling to the belt.

It is important to the understanding of a conveyor's (or a plant's) carryback problem to know how much residual material is carried on a given belt. When you quantify the problem–when you know how much carryback is present–you can determine how effective belt cleaning systems are, and what efforts–in the form of additional cleaning systems and improved cleaner maintenance schedules–are necessary.

The bulk material handling industry in Australia– both industrial operations per se and suppliers, consultants, and researchers serving the industry–has been a leader in designing and using systems to accurately assess the amounts of belt carryback. These procedures have been used to monitor the success of belt cleaning equipment and to aid in the design of equipment and performance of maintenance contracts on cleaning and conveying equipment.

To determine how much carryback is actually present, a small sample can be collected from a given portion of the belt. The sample can be obtained using a putty knife to manually scrape a portion of belt as soon as the belt is stopped.

Another method is to use a carryback gauge, composed of a sample collection pan affixed with scraper blades, to collect the material while the belt is running. Belt cleaning pioneer Dick Stahura is a longtime advocate of this style of collection device. **(Figure 13.4)** Great care must be used to safely collect material from a moving belt.

A further refinement of this system placed the collection pan in a track with a support mechanism **(Figure 13.5)** to remove the risk to the operator from the testing procedure.

Even these measurements are not without their limitations. Carryback is seldom uniform or equal across the belt, especially after one or several passes from belt cleaning systems. More likely it will be visible in bands or strips of varying density and width. As in any scientific measurement, it is important that the tests not be affected by the intent or ineptitude of operator.

The most accurate way to sample carryback is to use a gauge that samples the entire load-carrying width of the belt for a short period. Developed by Australia's TUNRA group (The University of Newcastle Research Associates, Ltd.), the ICT Carryback Gauge uses a moving sampler

Imperial Units		Metric Units
4 feet	Distance sampled	1200 mm
800 feet per minute	Belt speed	4 m/sec
0.005 minutes	Length of time for this amount of belt to pass point X.	0.005 minutes
0.035 ounces	Material collected (carryback on belt)	1 gram
Carryback on belt passing point X		
7.0 ounces per minute	Per minute	200 grams per minute
420 ounces per hour	Per hour	1200 grams per hour
26.25 pounds per hour	Larger units per hour	12 kilograms per hour
Assuming 5 day per week, 24 hours per day operation		
24 hours x 5 days = 120	Hours per week of operation	24 hours x 5 days = 120
3150 pounds per week	Carryback per week	1440 kilograms per week
1.575 tons per week	Larger units	1.4 metric tons per week
50 weeks	Weeks per year of operation	50 weeks
157,500 pounds per year	Annual carryback	72,000 kilograms per year
78.75 tons per year		72 metric tons per year

Figure 13.7

Annual Carryback Calculation.

that traverses the belt at a constant rate. **(Figure 13.6)** This approach uses a motorized system to drive a sampler back and forth across the belt at a constant speed.

There is the need to formulate and gain acceptance for standard procedures for sample collection and reporting that would allow industry-accepted standards to be developed.

Calculating the Carryback

By accepting this collected sample as the average amount of carryback (or better yet, actually scraping the belt at several points across its load-carrying width to achieve an average), it is possible to determine just how much carryback is present on the belt. To calculate this, multiply the sample weight by the belt's load-carrying width and the conveyor's operating schedule.

For example, the seemingly modest carryback amount of 0.035 ounces (one gram)–removed from a section of belting four feet (1.2 m) in length–shows that there can be carryback material weighing 1.575 tons (1.4 metric tons) per week on this single conveyor. **Figure 13.7** shows the calculation in both imperial and metric units.

One gram–the amount of material collected from four feet of conveyor–is the same amount of material as the small packets of sweetener found on the table in most restaurants. Yet if this amount of material is left on the belt, and the plant's schedule is extended to around-the-clock operations, seven-days-per-week, the carryback total grows to more than 110 tons (100 metric tons) of material that can be left on the belt per year.

The 0.035 ounces (one gram) of material per four feet (1.2 m) of belt length used in the above examples represents so small a quantity of material that some would wonder why worry about this belt at all; it is already clean. More typically, actual measurement of material on many belts shows carryback in the neighborhood of 1/4 to 1/2 ounce (7 to 15 grams) per four-foot (1.2 m) sample. This is material that is carried back on the belt and can be released from the belt surface, becoming airborne dust and/or accumulating in piles under the conveyor or in buildups around rolling components.

A more "scientific" measurement would be to sample the belt right after the discharge point and then scrape a portion of the belt just prior to the loading zone. In addition to showing how much material adheres to the belt past the discharge, this would yield a measurement of how much carryback is dropped from the belt during the return run.

Cleaner Performance Monitoring

The use of carryback measurement systems will allow the development of a belt cleaning performance specification for a given material handling facility. This specification will detail the design, supply, installation, commissioning, and maintenance of belt cleaning systems (subject to compliance with performance guaranteed carryback levels).

After belt cleaning system installation, tests are conducted to assess the performance of the cleaners. The results provide information for use in "fine tuning" the system and evaluating the need for additional or alternative systems. A continuing testing program and a database will also yield information on the requirements and payback for cleaning system maintenance.

This monitoring allows the facility to assess the savings to be made through upgrading to more efficient cleaning systems.

This performance analysis and maintenance program could be implemented by in-house departments or through the awarding of a service contract to a specialist in the supply, installation, maintenance, and analysis of belt cleaning systems.

Designing Conveyors for Effective Cleaning

When considering the construction of new conveyors, it is desirable to include in the design requirements a specification for belt cleaner performance. This could be in the form of a specified amount of carryback actually allowed to pass the cleaning system. This standard might encourage plant owners to demand (and engineers to design) conveyor systems with adequate cleaners to hold carryback to this specified level. In addition, it would encourage the use of other systems and components on the conveyor that are compatible with the aim of effective belt cleaning.

The most common problem in the employment of belt cleaning devices occurs when there is insufficient room provided in the design of the head frame and housing for an adequate multiple belt cleaner system. This commonly occurs because conveyor engineers have not taken into account the true nature of the conveyed material, particularly when it is in its worst condition.

One of the keys to effective belt cleaning is the mating of blade to belt. It stands to reason that the more perfectly the blade matches the belt profile, the better it will clean that profile. Anything that makes it more difficult for the blade to stay in perfect contact as the belt moves must be avoided in the design of a conveyor system. These factors would include the wing pulleys, out-of-round pulleys, and poorly-selected or poorly-installed lagging. Any pulsation or vibration of the belt's surface can lower cleaning efficiency and adversely affect the life of the belt.

Cleaning is improved by the use of vulcanized splices on the belt. Improperly installed mechanical fasteners can catch on cleaners and cause them to jump and vibrate or "chatter." If mechanical splices are used, they should be recessed to the supplier's recommendations to avoid unnecessary damage to the cleaner and the splice itself.

After a belt cleaner is installed, periodic inspection, adjustment, and maintenance will be required. Just as the cleaners must be designed for durability and simple maintenance, the conveyor must be designed to allow this service, including the required access and clearances.

Cleaning Systems for "Worst Case" Materials

Conveyor cleaning systems should not be designed only to match the limited challenges of "normal" operating conditions. Rather, they should be designed to stand up to the worst possible applications that may be encountered. Even if 99 percent of the time the conveyed material is dry, the day will surely come when the material will suddenly be very wet and sticky. If there is a bad run of material, and the cleaning system is not adequate for the challenge, it may take just a matter of minutes for the conveyor line to become a mess of mistracking belt, piles of fugitive material, and coated idlers. This will undo all the time, effort, and funds expended in the design and maintenance of the cleaning system and perhaps of the entire conveyor.

If the cleaning system is designed to the requirements of the "worst case," it will be a case of "overkill" during normal operating conditions. But this over design is likely to provide the benefit of improved wear life and reduced service requirements when operating conditions are normal. And when the conditions change for the worse, the cleaning system will be able to stand up to the challenge.

To analyze the material in its worst condition may take some work, but it is worth the effort. It must be done to have an efficient belt cleaning system under all conditions.

Analyzing the Material

The behavior of bulk material on a belt depends on a number of parameters, including bulk density, interface friction, interface cohesion, interface adhesion, and belt condition. Each of these parameters is dependent on previous loading and unloading of the material, the particle size of the material, and the moisture content of the material. The bulk density of the material should be tested at a number of moisture contents and simulated pressures. The pressure of the material against the belt is estimated using the following factors: the amount of material conveyed, previous loading and unloading, amount of moisture, the distance material is conveyed, and speed of the conveyor.

The particle size range of the material, while important in some handling aspects, is not generally an important factor in belt cleaning. During transport, the finer fractions of the material will migrate to the belt surface through vibration and impact. Bulk material testing is performed only on this fine fraction, while larger particles are generally ignored.

The most dominant factor affecting amounts of carryback and belt cleaning efficiency is the moisture content of the fine portion of the material conveyed. Testing has shown that for most (if not all) materials, adhesion to the belt increases as moisture increases to a maximum point, where adhesion then decreases. As shown in "bell shaped" curves graphed in **Figure 13.8**, an increase in moisture increases material cohesion to a maximum point, where it then decreases. The exact variation in adhesion with moisture content will vary from material to material and site to site.

Testing to determine the adhesion of a material to the belt can be omitted, as it can be found mathematically from the values of friction and cohesion. An optimal cleaning pressure can be determined using these results and belt cleaner specifics (cleaning angle, blade profile, and blade composition). Advanced modeling techniques can be used to predict the blade-to-belt pressure necessary to remove the carryback layer. In practice, a number of tests are performed at different moisture contents to determine a "worst case" operating scenario; that is, the highest pressure that will be required under the broad range of operating conditions.

The important thing to remember is that eventually the material will see the "worse case" during the life of the belt.

Universal Belt Cleaners

A quick review of the marketplace shows there are many competing designs for cleaning systems. This poses the question: How come industry has not settled on one successful design that can provide acceptable results in all applications?

An engineer for a major mining operation wrote in a summary of belt cleaners at his operations, that

> *"Due to the diverse materials with their greatly differing physical properties as well as the diverse environmental conditions in open cast mining, there is currently no universally applicable belt cleaner, which would fulfill the requirements of all situations without problems."*

And that writer was discussing only one company's German open cast lignite mines. That is a challenging application to be sure, but it is only one of the myriad of environments required for belt cleaners. The universe of belt cleaner applications is so much broader–different materials, different conditions, different conveying equipment–that it requires a number of different choices.

The problem is, of course, to provide a cleaning system that is adjustable to fit most of these situations.

Indeed, one of the problems in developing a universal cleaning system is that each manufacturer, each engineer, and plenty of plant maintenance personnel have their own ideas on what would work. And because there are so many variables in plants, equipment, materials, and conditions, all of these designs have found some application and some level of success.

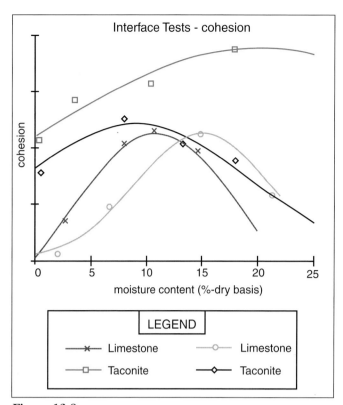

Figure 13.8
Testing shows that raising the moisture content increases cohesion to a certain point for each material, after which cohesion decreases.

Suppose a plant engineer or maintenance worker were to design his own belt cleaner and then have it installed. He would stop by his cleaner every day to see how it was working. He would pay attention to it, periodically adjusting the level of blade-to-belt tension and knocking off the material buildup. Given this level of attention, chances are pretty good that this cleaning system would provide performance that was at least equal to many of the commercial cleaning systems presently available. But rather than the device's design, the key element has been the level of service that allowed the system to maintain its cleaning performance.

The Basics of Effective Cleaning Systems

The basic "wish list" for the installation of an effective belt cleaning system would include the following criteria:

Far Forward

To minimize the release of carryback into the plant environment, belt cleaning should take place as far forward on the conveyor return as possible. Residual material should be removed by cleaners positioned near the conveyor discharge. Typically, at least one cleaner is installed against the belt while the belt is still in contact with the head pulley. This position provides two advantages: it minimizes the distance carryback is carried (reducing the potential for release onto rolling components and into the plant environment) and by supporting the blades against a firm surface, it encourages more effective cleaning.

If the cleaners are tensioned against the belt while the belt is still against the head pulley, control of blade-to-belt pressure can be more precise. If the belt is not held against a pulley, fluctuations in the belt's travel and tension will affect cleaning pressure, alternatively bouncing the blade onto and off the belt ("chatter") and/or allowing the blade to penetrate the belt too deeply and cause damage.

With cleaners installed forward, it is easier to redirect the removed carryback into the main material flow. This reduces the chance for material to escape and, likewise, reduces the need for dribble chutes or other mechanisms to scavenge material.

In addition, the far forward position for the initial cleaner allows the most possible space for the installation of one or more secondary cleaners. As with the pre-cleaner, the farther forward each secondary cleaner is, the less chance there is for the carryback to be released into the plant environment.

Out of the Material Flow

It is important that cleaners be installed out of the flow of the main material body and that the cleaned materials do not adhere to the blades and structure.

If the cleaner is installed in the material trajectory, the mainframe and the back of the blades will wear out from the bombardment of discharging material before the cleaning edge wears out. Proper placement usually means installing a pre-cleaner so the blade tip is below the horizontal centerline of the pulley. This position will increase the effectiveness of the cleaner and extend the life of blades and mainframe.

Cleaners should be installed outside of the material trajectory. However, they will still acquire a buildup of material that adheres to their outside surfaces and should, therefore, be designed to minimize the chance for material adhesion. This should be accomplished through the avoidance of flat surfaces or pockets that can catch materials and through the use of non-stick materials for cleaner construction. In the proper environment, water can be sprayed on the surface of the belt, or even on the cleaners themselves, to assist in minimizing material buildup.

With Minimum Risk to the Belt

An essential consideration in the selection of any belt cleaning device is minimization of any risk that the cleaners themselves could damage the belt or a splice–the very systems they were installed to protect. A belt cleaner must be designed to prevent damage and improve operating efficiency; it is counterproductive for a cleaner to possess even the remotest possibility of damaging a belt.

The cleaning system must be designed so it can flex away from the belt. This will ensure safety when a mechanical splice or a section of the belt that is damaged (or even repaired) moves past the cleaning edge. The cleaner's tensioning systems–particularly on the pre-cleaner, where the angle of attack is more acute–must include a mechanism to provide relief from shock of splice impact. The pre-cleaner blades must be designed to allow for easy splice passage.

If the life expectancy of the belt is short–say, six months, as opposed to the more typical two-to three-year life span–it might be worthwhile to be more aggressive with the cleaning system. For example, the use of tungsten carbide blades on the pre-cleaner would provide more effective cleaning. Over time, these blades pose a greater risk of belt wear and damage. However, a risk/reward analysis shows they could improve cleaning performance without significantly increasing belt replacement costs. The belt is going to be replaced anyway, so why not get superior cleaning performance.

Strips of used belting should never be applied for belt cleaning, because these strips may include steel cables or abrasive fines in the surface of the used belting. These embedded materials can cause excessive wear in the running belt's top cover.

The Givens of Cleaner Design

Blade cleaners are passive systems; they require no drive mechanism (electricity, water, or compressed air) in order to function (although they require an energy source–a spring, compressed air reservoir, or twisted rubber band to hold the cleaning edge against the belt). The blade element that directly contacts the belt is subject to abrasive wear and must be readjusted or replaced at installation-specific time intervals to maintain cleaning performance.

Typically, cleaners are made with blades that are not the full width of the belt. For example, a 72-inch belt will use a cleaner with a blade that is 66 inches wide. An 1800 mm belt will use a cleaner with coverage of approximately 1650 mm of belt. The difference in width does not affect cleaning performance, because the full width of the belt is not typically used to carry material. It also allows minor variations in belt tracking and provides a better fit on troughed belts.

Testing indicates that a cleaning edge composed of narrow, independently-suspended blades will stay against the belt's surface a higher percentage of the time than a single "slab" blade. A single blade must be constructed to resist the forces of its total surface contact. A multiple-blade design with individual spring or elastomer support will keep each blade in proper cleaning tension against the belt, yet allow each individual blade to yield to a lower pressure than the

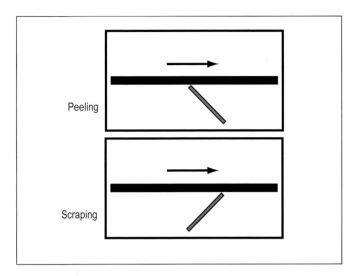

Figure 13.9
Alternative Angles of Belt Cleaning.

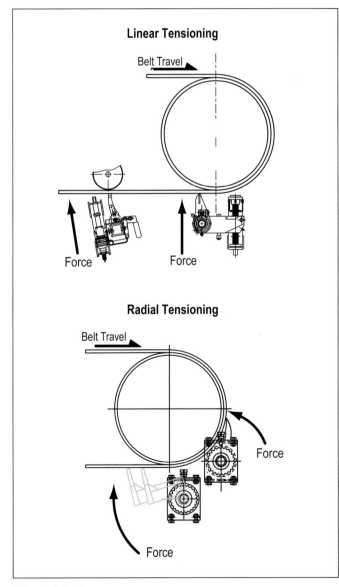

Figure 13.10
Alternative Ways to Adjust the Cleaner to the Belt.

tensioning device's total applied force. In other words, narrow blades can match up better against the belt, follow changes in surface contour, bounce away from the belt for splice passage, and return more easily than a single, continuous blade. This means a proper multiple-blade design is more efficient and safer for the cleaner, the belt, and the operator.

There are a number of materials used as cleaner blades. They range from rubber and urethane to mild and stainless steels. In addition, blades are available with inserts of tungsten carbide or fillers such as glass beads to enhance abrasion resistance and cleaning performance. It is important to note that if the material held against the belt is harder than the belt, the risk of belt damage increases, as does the rate of wear of the belt's top cover. Cleaner manufacturers have extended the range of urethane materials available to provide improved performance for specific conditions, including improved resistance to abrasion, heat, chemicals, or humidity. In some cases, the unique combination of characteristics in a specific application require a comparative testing program to determine the best material for that application.

Peeling vs. Scraping

The angle of attack for the cleaning blades against the belt is a subject of importance. Generally speaking, there are two alternatives: peeling blades and scraping blades. (**Figure 13.9**) With peeling, the blades are opposed to the direction of belt travel; in scraping, the blades are inclined in the direction of travel, typically at an angle of 3 to 15 degrees from the vertical. Each design has its advocates.

Metal blades in a peeling position are quickly honed to razor sharpness by the moving belt and can do expensive damage if they are knocked out of alignment. Peeling blades are also subject to high-frequency vibration that causes the blades to "chatter," repeatedly jabbing the sharpened edges into the belt cover.

Scraping blades allow material to build up on the inclined cleaning edge, which can pull the blade away from effective cleaning contact. With a scraping blade, the upstream edges of the cleaning blade will not bite into the belt surface, even if held against the belt with excess pressure.

A general opinion is that the peeling angle is acceptable for primary cleaners, which are applied at very low pressures against the belt due to their position and their construction. However, it is advisable to use blades in a scraping position in secondary cleaners, where higher blade-to-belt cleaning pressures and metal blades present more risk to the belt, splice, and cleaner itself.

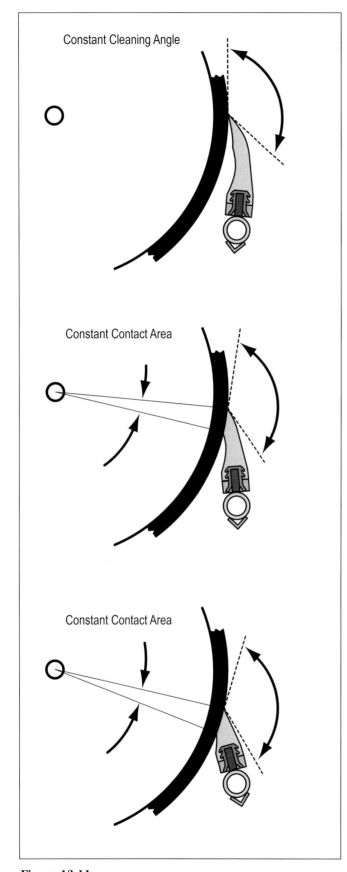

Figure 13.11

With a "Constant Angle" blade design, the cleaning angle remains the same throughout the life of the blade.

Linear or Radial Adjustment

There are two competing theories for belt cleaner adjustment. There are linear-adjusted cleaners that are pushed up (in a line) against the belt and radially adjusted cleaners that are installed with a mainframe as an axis and rotated into position. **(Figure 13.10)** Radially-adjusted cleaners have several practical advantages over the linear design. They are easier to install, can be adjusted from one side of the belt, and can more readily rotate away from the belt to absorb the shock inherent in belt motion and splice passage.

However, linear-adjusted cleaners have had one advantage: the cleaning angle–the angle at which the blade sits against the belt–remains constant. The blade remains positioned at the same cleaning angle, regardless of the state of blade wear. Without prejudice to which type of cleaner is the best choice, it is obvious that a radially-adjusted cleaner can work at maximum efficiency for only a portion of its blade life.

Maintaining the angle of the blades against the belt is important for ensuring effective cleaning. If the angle of contact is altered by blade wear, cleaner performance will similarly "decay." A well-designed belt cleaner must control the cleaning angle across its wear life.

Constant Angle Cleaning

To overcome the problem with changing blade angle, one design in radially-adjusted belt cleaners incorporates a curved blade. This design is termed "CARP", for Constant Angle Radial Pressure. Designs incorporating this CARP technology feature a curved blade designed so as the blade wears, the blade maintains the same angle against the belt. **(Figure 13.11)** This design has the obvious advantage of maintaining cleaner efficiency. It also eliminates such problems as vibration or "chatter" stemming from incorrect cleaning angles and the over tensioning of worn cleaner blades.

Constant Area Cleaning

A new cleaner blade has a small contact area on the belt. Blades are usually designed with a "point" to allow them to "wear in" quickly to achieve a good fit to the belt, regardless of the head pulley diameter. Typically, as cleaner blades wear, the surface area of the blade touching the belt increases. This causes cleaner efficiency to decline because of the relative reduction in blade-to-belt pressure. Therefore, the system's tensioner requires adjustment (retensioning) to provide the additional pressure per square inch of belt contact area for consistent cleaning performance. It would be better to design cleaners that do not suffer from this gradual increase of blade-to-belt area. The CARP principle cited above has also proved capable of minimizing the change in area throughout a blade's wear life.

Optimal Cleaning Pressure

A key factor in the performance of any cleaning system is the ability to sustain the force required to keep the cleaning edge against the belt. Blade-to-belt pressure must be controlled to achieve optimal cleaning with a minimal rate of blade wear. Many people have the misconception that the harder you tension the cleaner against the belt, the better it will clean. Research has demonstrated this is not true.

A 1989 study by the Twin Cities Research Center of the U.S. Bureau of Mines examined the issue of what blade-to-belt pressure will offer the best level of cleaning, without increasing blade wear, belt damage, and/or conveyor power requirements. This research is published in *Basic Parameters of Conveyor Belt Cleaning*, by C.A. Rhoades, T.L. Hebble, and S.G. Grannes (United States Department of the Interior Bureau of Mines Report of Investigations 9221). The study evaluated cleaning effectiveness and wear characteristics of various steel blades by holding them perpendicular to a running belt with measured amounts of pressure to remove a moistened sand/lime mixture.

The Bureau Of Mines' study found that the amount of both carryback and blade wear decrease with increasing blade pressure until an optimum blade pressure is reached. The study established this optimum blade-to-belt pressure at 11 to 14 psi (76 to 97 kPa) of blade-to-belt contact. **(Figure 13.12)** Increasing pressure beyond the 11 to 14 psi (76 to 97 kPa) range raises blade-to-belt friction, thus shortening blade life, increasing belt wear, and increasing power consumption without providing any improvement in cleaning performance. An overtensioned blade will usually show flat, even wear, and may also show some discoloration such as "scorch" marks.

Less pressure causes less effective cleaning and can also cause rapid blade wear. A belt cleaner can lay loose against the belt and appear to be cleaning, while in reality, material is being forced between the blade and the belt at high velocity. This passage of material between the belt and the blade creates channels of uneven wear on the face of the blade. As material continues to pass between the blade and the belt, these channels increase in size, rapidly wearing the blade. A blade that is undertensioned will usually show a jagged edge with wear lines on the wear surface.

The Bureau of Mines' study also reported it found that cleaning effectiveness decreased over time because of uneven blade wear. Grooves worn into the blade edge allow the passage of carryback that cannot be eliminated by increasing the blade-to-belt contact pressure. The Bureau of Mines' report notes, "Once a cleaner blade's surface is damaged, no realistic level of blade-to-belt pressure will allow the blade to conform to the belt's surface for proper cleaning."

Tensioning Systems

Blade-to-belt cleaning pressure is maintained by some form of tensioning device. Tensioners range in sophistication from concrete block counterweights and locking collars to torque storage couplings and engineered air spring systems plumbed to the plant's compressed air supply. The choice for a given installation depends on conveyor and cleaner specifications and plant preferences.

All tensioning systems should be designed to allow the cleaning edge to relieve itself away from the belt to allow passage of mechanical splices and other obstructions. Tensioners should also be self-relieving to minimize risk of injury to personnel or equipment if the blades are

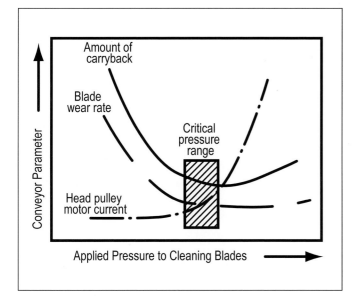

Figure 13.12

Effect of Blade to Belt Pressure on Cleaning Efficiency and Blade Life, as shown in testing by the US Bureau of Mines.

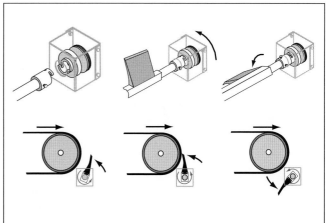

Figure 13.13

To minimize the risk of damage or injury, tensioners should self-relieve, automatically uncoupling if the blades "pull through."

"pulled through" by obstructions or holes in the belt. **(Figure 13.13)**

When adjustment and retensioning are required, the cleaner and tensioner unit should allow this maintenance to be performed simply, without needing tools or more than one service worker.

One recent innovation in cleaning technology uses the resilience of urethane blades when compressed and locked into position to supply the cleaning pressure. **(Figure 13.14)** When installed, these blades are deflected by being forced against the belt. As the blade wears, it "stands taller" to maintain cleaning pressure. Because the blade itself supplies both cleaning pressure and shock-absorbing capacity, the cleaner does not need a conventional tensioner. Instead, the blade assembly is forced against the belt, and the mainframe is locked into position. **(Figure 13.15)**

A similar innovation includes the addition of "secret springs" molded inside urethane blades. **(Figure 13.16)** When tensioned into cleaning position, the urethane blades flex against the belt, storing cleaning pressure. This pre-loading allows the natural resilience of the urethane to combine with the force of the "secret spring," molded inside each modular blade, to apply consistent pressure across the life of the blade. This consistent cleaning pressure works to virtually eliminate the need to retension the cleaner to maintain cleaning performance over the life of the blade. The "secret springs" provide the cleaner with a variable wear rate, allowing the cleaner to compensate for the high levels of abrasive wear in the middle of the belt, while maintaining consistent cleaning pressure across the belt. **(Figure 13.17)**

Specifying a Belt Cleaning System

Selecting a belt cleaner for a given application requires the assessment of a number of factors. The following is the basic information that a supplier would need to know to provide a suitable belt cleaning system.

- Belt Width and Speed (including Actual Width of Cargo on Belt)
- Pulley Diameter
- Material Characteristics (including Moisture Content, Temperature, Abrasiveness, and Corrosiveness)
- Conveyor Length
  Conveyor length is a significant variable, as the undulating action of the belt as it moves over the idlers causes the fines to settle through the material and become compacted on the belt. On long overland conveyors, this effect can be significant. For this reason, longer conveyors are almost always harder to clean than shorter belts.

Additional variables that may affect the ultimate performance of the selected system (and that, therefore, should be reviewed in the selection of a cleaning system) include:

- Changing material characteristics (i.e. wet and sticky to dry and dusty)
- Severe temperature extremes
- Cuts, gouges, rips, or cracks in the belt surface resulting from age or abuse
- Numerous, non-recessed, or damaged mechanical splices in the belt surface
- Material buildup on head pulleys and other rolling components causing belt vibration, making it difficult to keep a cleaner in contact with the belt
- Material adhering to the cleaning device
- Material accumulation in the dribble chute, encapsulating the cleaners

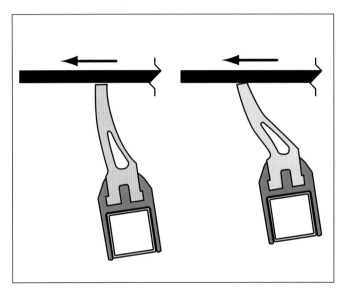

Figure 13.14
Flexed urethane blades use the resilience of the urethane to supply cleaning pressure.

Figure 13.15
By deflecting the blades against the belt, this cleaner can self-adjust to compensate for wear.

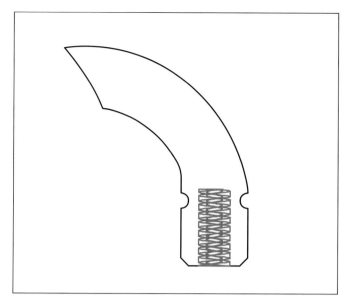

Figure 13.16
A spring molded inside this urethane blade allows the cleaner to maintain cleaning pressure across the life of the blade.

Figure 13.17
With the spring molded inside, this blade achieves a variable wear rate that compensates for the higher abrasion in the middle of the belt.

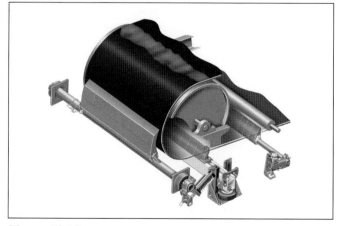

Figure 13.18
A multiple cleaning system includes a pre-cleaner on the face of the head pulley and one or more secondary cleaners.

Other considerations in the development of a supplier proposal and the evaluation of that proposal include:
- Level of Cleaning Performance Desired/Required
- Physical Space Available
- Level of Maintenance Required/Available
- Level of Installation Expertise Required/Available
- Initial Price vs. Cost of Ownership
- Manufacturer's Record (Service Capabilities and Performance Guarantees)

It is worthy of consideration that the removal of the last little bit of carryback from a belt will require significant effort and expense in cleaning systems and maintenance labor. There are very few situations that justify specifying a system that will provide a belt that is perfectly "white glove" clean.

It can be estimated that 50 percent of the carryback that remains on the belt past the cleaning systems will stay on the belt throughout the conveyor's return run. The material will still be on the belt when the conveyor enters the load zone again.

Keep in mind that if the carryback material does not fall off the belt, it does not need to be cleaned off; it may be wise to specify a lower standard of cleaning. The cleaning system should be designed, installed, and maintained to meet actual operating requirements, not some artificially high goal.

The Systems Approach to Belt Cleaning

The state-of-the-art in cleaning technology understands that more than one "pass" at the belt must be made to effectively remove carryback material. Like shaving, it is more effective to make a series of gentle strokes over the surface, than an "all at once," high-pressure assault.

It is usually most effective to install a multiple-cleaning system, composed of a pre-cleaner and one or more secondary cleaners. (**Figure 13.18**) The pre-cleaner is installed as a doctor blade, using low blade-to-belt pressure to remove the top layer (and majority) of the carryback. The secondary cleaner(s), tensioned at the optimum belt-cleaning pressure, to perform the final precision or mop-up removal of adhering fines without being overloaded with the mass of carryback. The two styles of cleaners are given different responsibilities in the task of cleaning the belt and so are designed and constructed differently. In addition to providing improved cleaning, a multiple-cleaner system increases the maintenance interval—the time between required service sessions.

The phrase multiple cleaning systems could refer to any combination, ranging from the commonly seen pre-cleaner and secondary cleaner system (**Figure 13.19**) to more sophisticated systems that include a pre-cleaner and one or several secondary cleaners, and/or a tertiary cleaner of yet another type. These could possibly be combined with a belt "wash box," incorporating sprayed water and wipers for "squeegee" belt drying.

Secondary cleaners are inefficient when they have to cope with coarse grit because their negative angle of attack of the blade to the belt–chosen to reduce the risk of belt damage–forces this coarse material into the pinch point at the junction of the blade and belt. This grit is then forced over the blade edge, opening a gap between the blade and the belt. The gap allows a continuous stream of grit through, and no amount of pressure will effectively close it. If sufficient pressure is initially applied to prevent this gap from occurring, then the cleaning force is enough to dramatically injure the belt cover, especially if the conveyor runs dry and empty.

Pre-cleaners, with low blade pressure but an aggressive angle of attack, are able to shear this coarse layer of grit off the bottom layer of fines. This clears away the top layer and enables the secondary cleaner to remove additional material. Of course, this same process still occurs in microform, with each successive secondary cleaner removing material exposed by an earlier cleaner. This is the reason that each successive cleaner always seems to take off some material, and perfection is seldom reached.

Figure 13.19
The pre-cleaner and secondary cleaner have different responsibilities and so are designed and constructed differently.

Pre-Cleaners

A pre-cleaner functions as a metering blade to shear off most of the carryback from the belt, leaving behind only a thin skim of fines. This controls the amount of carryback that reaches the secondary cleaner(s), protecting them from being overloaded by a flood of material.

The pre-cleaner, sometimes called the primary cleaner or the doctor blade, is installed on the face of the head pulley just below the trajectory of the material discharging from the belt. **(Figure 13.20)** This position allows the material removed from the belt to fall in with the main cargo as it leaves the belt, minimizing the overloading of the dribble chute or other reclamation system.

This places the pre-cleaner blade in a peeling position; it is inclined against the movement of the belt and pulley at an angle between 30 to 45 degrees. Using this low angle of attack, in combination with elastomer pre-cleaner blades applied with light pressure against the belt, results in low wear rates for both the blade and the belt surface. If the angle of attack were greater (i.e., a scraping position), more pressure would be required to hold the blade in position against the onslaught of material. Increasing this pressure would increase the risk of damage. Rather than blocking the path of the material, the pre-cleaner diverts the material away from the belt, so it can return to the main material flow or move down the back of the cleaner into the discharge.

To minimize the risk to belt, splice, and cleaner from even a lightly tensioned blade in a peeling position, pre-cleaners should always use resilient urethane or rubber blades (rather than metal) and be only lightly tensioned against the belt. A blade-to-belt pressure of approximately 2 psi (13.8 kPa) of blade material on the belt combines belt safety with cleaning performance. This low blade-to-belt pressure means that the tensioning system will be able to relieve–to bounce the blade away from the belt–when an obstruction such as a

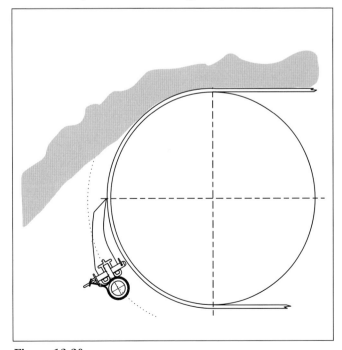

Figure 13.20
The pre-cleaner is installed on the face of the head pulley just below the material trajectory.

mechanical splice moves past the cleaning edge. This reduces the risk of damage. The lower cleaning pressure also improves blade life and reduces belt wear.

There is a temptation to attempt complete cleaning with a single pre-cleaner. To accomplish this, the cleaner is tensioned against the belt with higher blade-to-belt pressure than would be allowed by respect for preserving the belt, splice, and cleaner. Multiple blades lightly tensioned against the belt are far more effective and kinder to the belt than a single, "slab" cleaning edge forced against the belt with high pressure.

Secondary Cleaners

Secondary cleaners are designed to remove the remainder of material–the sticky fines that have passed under the blade of the pre-cleaner. If the pre-cleaner is installed to perform the initial cleaning, the secondary cleaner is dedicated to performing the final "mop-up" to prevent the escape of fugitive material. (More than one secondary cleaner may be required to achieve the desired level of cleanliness).

The positioning of the secondary cleaner(s) is important. The closer the removal of carryback is performed to the conveyor's discharge point, the lower the risk of problems with the buildup of fines in the dribble chute. It is best to place the secondary cleaner blades in contact with the belt while the belt is still against the head pulley. **(Figure 13.21)** This allows the secondary cleaner (or the initial secondary cleaner, if more than one are used) to scrape against a firm surface for more effective material removal. If additional secondary cleaners are installed, or due to conveyor structural members or space limitations, it is impossible to install a cleaner in this preferred position, the cleaner(s) should be mounted to clean against snub pulleys, return rollers, or other components that give a firm profile to the belt. If a secondary cleaner is installed in a position where its pressure against the belt changes the belt's line of travel **(Figure 13.22)**, cleaning performance will be ineffective. In this case, increasing the applied pressure will only serve to change the line more and wear components faster, without improving cleaning performance.

The angle of attack of the blade to the belt is also an important consideration. Blades in a peeling position are quickly honed by belt movement to extreme sharpness. These sharpened blades raise the risk an untrained or "in a hurry" adjustment will apply too much pressure to allow instant release from belt obstructions, such as mechanical splices. The result could be damage to the belt, splice, or the belt cleaner itself. Consequently, secondary blades should be angled in the direction of the belt's travel–a scraping angle–rather than opposing the belt in a peeling position. Testing has indicated that an angle of 7 to 15

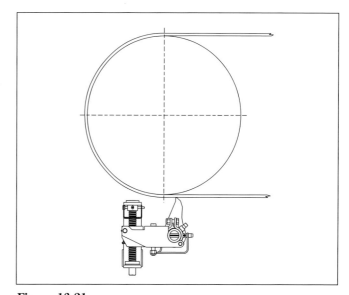

Figure 13.21
The ideal location for the secondary cleaner is at the point where the belt is leaving the head pulley.

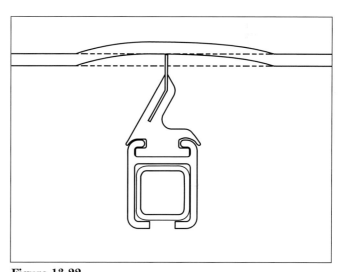

Figure 13.22
Cleaning performance will be reduced if the cleaner is installed where it can alter the belt's line of travel.

Figure 13.23
Belt cleaners with overlapping blades will eliminate the "stripes" of carryback allowed by gaps in the cleaning edge.

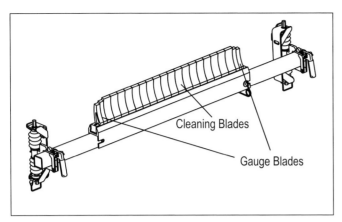

Figure 13.24
Flexed urethane allows this belt cleaner to self adjust to maintain cleaning pressure. Shorter blades at each end serve as wear indicators.

Figure 13.25
Oversize belt cleaning systems have been developed for improved cleaning and reduced maintenance in tough applications.

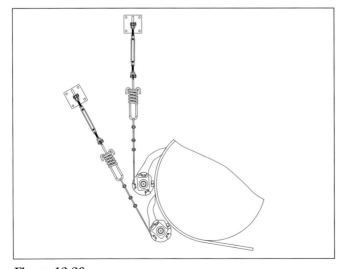

Figure 13.26
The large head pulley of this high-speed conveyor allowed installation of two pre-cleaners.

degrees in the direction of belt travel maintains cleaning efficiency while allowing easier splice passage.

Because this cleaner must be as efficient as possible, the blade must mate evenly with the belt. The moving belt does not present a consistent and uniform surface. There are peaks and valleys that the cleaner must adjust to instantly. Narrow, independent blades that are individually suspended have the best potential to remain in precise contact as the belt surface passes across the cleaning edge. It also helps if the individual blades can pivot or rock from side to side to instantly adjust to the changing contours of the belt surface. Research indicates that a design using multiple blades approximately six to eight inches (150 to 200 mm) wide is well suited for effective cleaning.

The Bureau Of Mines' study *Basic Parameters of Conveyor Belt Cleaning* points out that initial blade wear occurs at the edges where blades adjoin. Their testing showed material would go through the adjacent space between the blades, and these spaces slowly enlarge, accelerating blade wear and allowing more material to pass. To minimize this erosive wear, an overlapping blade pattern can be used, created by an alternating long arm/short arm pattern. **(Figure 13.23)** This prevents "stripes" of carryback down the belt created by gaps between the blades.

The blades themselves can be of a hard material–steel or tungsten carbide, for example–that resists the buildup of heat stemming from the friction against the belt's surface. Some operations prefer to avoid application of a metal blade against the belt, and so a variety of urethane formulations are available.

A recently developed cleaner applies the natural resilience or "spring" of the urethane used in the blades to reduce cleaner adjustment. The blades in this new cleaner are forced against the belt, so they "flex" in the direction of belt travel. **(Figure 13.24)** Over time the blade's resilience pushes them up against the belt, even as they are being worn away by the movement of belt and material. This reduces the need for maintenance and adjustment, as the blades are now self-adjusting. Installed as a secondary cleaner, this "set and forget" system is locked into position, yet maintains cleaning pressure without service attention.

Matching Cleaner to Application

The increasing sophistication of belt cleaner design has allowed the development of cleaners to conform to the needs of each application. These alternatives include specialized materials for cleaner construction, such as high-temperature, high-moisture, and high-abrasion urethanes. In addition, there now are cleaners engineered to match specialized applications ranging from light-duty food-grade to heavy-duty mining applications.

Mine-Grade Belt Cleaners

The high volumes of material and the fast conveyors, wide belts, and large pulley diameters used in many mining operations (especially surface mines) pose special challenges for belt cleaning systems. The removal of overburden in some German lignite mines employs conveyors with belt widths up to 126 inches (3200 mm) wide, operating at speeds of 2067 feet per minute (10.5 m/sec). Extra-heavy-duty mine-grade systems have been developed to meet such needs. **(Figure 13.25)** These systems are marked by massive I-beam mainframes to withstand the high volumes of material and large lumps, massive blades to provide extended wear life, and durable tensioning system to reduce the need for system maintenance (and the downtime required to perform the maintenance).

The high speed of these conveyors makes it difficult to use secondary cleaners effectively. The higher operating speeds (and resultant higher vibration) of these belts combine with the higher blade-to-belt pressure of a typical secondary cleaner to produce both higher wear and added risk to both belt and cleaner. As a result, many of these applications feature the installation of two pre-cleaners on the head pulley; fortunately, the pulleys are sufficiently large enough to allow this practice. **(Figure 13.26)**

Cleaners for Reversing Belts

Some conveyors run in two directions or have substantial rollback, so it is critical that cleaners work well in either direction of operation (or at least are not damaged by belt reversal). Specialized cleaners have been developed for reversing belts. These cleaners are installed so they are tensioned vertically against the belt. Typically, the cleaners are designed with a vertical blade that deflects a modest amount–7 to 15 degrees–under belt motion in either direction. **(Figure 13.27)**

Of course these reversing cleaners can be used on one-direction belts. A feature of these reversing belt cleaners that encourages their use on non-reversing belts is their vertical installation and tensioning. This allows the cleaner to be fit into narrow spaces where larger cleaners with a trailing arm design would not fit.

Food-Grade Cleaning Systems

These systems are constructed of food-grade materials and engineered for the small pulleys and slower belt speeds common to food-processing plants.
(Figure 13.28)

Chevron Belt Cleaners

Belts with ribs, cleats or chevrons are used to convey materials that would fall back as the belt moves up an incline, but the ribs that contain the material pose problems during the removal of carryback. Chevron belt cleaners use blades with "fingers" that are able to walk over the obstructions. This design can effectively clean chevrons/ribs/cleats up to 3/4 inch (20 mm) tall.
(Figure 13.29)

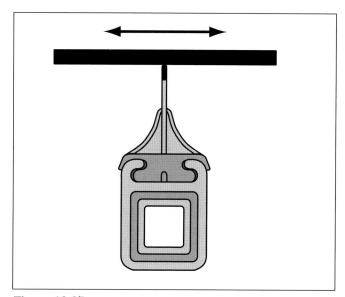

Figure 13.27
The reversing belt cleaner is suitable for installation on belts that run in two directions.

Figure 13.28
A compact food-grade belt cleaner is designed for applications on the small belts typical of food processing operations.

Figure 13.29
The chevron belt cleaner features blades with "fingers" that "walk over" obstructions on the belt.

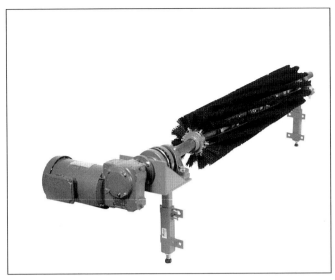

Figure 13.30
A powered rotary brush cleaner is best suited for dry material.

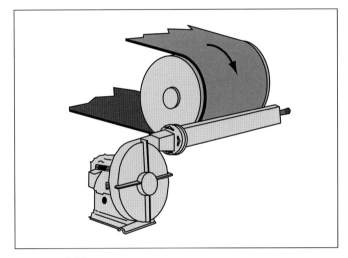

Figure 13.31
The pneumatic belt cleaner uses air supplied by fan to remove material from the belt.

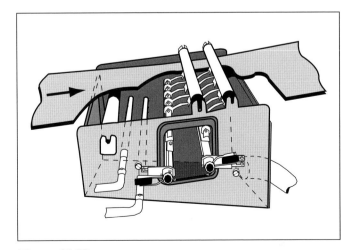

Figure 13.32
A "wash box" incorporates both water spray and conventional belt cleaners.

Rotary Brush Cleaners

These cleaners could be free-wheeling (turned by the motion of the belt) or driven by an electric motor. They can be effective on dry materials; however, they typically encounter problems with sticky or moist material that can build up in the bristles of the brush. **(Figure 13.30)**

Pneumatic (Air Knife) Cleaners

These direct a stream of air over the face of the pulley to shear off carryback. This system can be effective in removing dry materials; their disadvantages include the continuing expense of providing the air stream and the creation of additional airborne dust that will settle someplace in the plant. **(Figure 13.31)**

The Use of Water in Belt Cleaning

In many applications an increase in moisture level in the material is directly related to increased difficulty in handling and removing the material. In other cases, the material specification strictly limits the moisture level, and any increase in moisture results in financial penalties. However, the use of water is a major advantage in cleaning conveyor belts handling almost any material.

Sticky material will adhere to the belt and anything else it contacts. These materials can adhere to belt cleaner blades as easily as to any other part of the system. A belt cleaner blade generates heat through friction as it rubs against the belt surface. The sticky material not only adheres to the blade, it can actually be baked on. This baked-on buildup will allow further material to accumulate, and the problem continues and accelerates.

A small amount of water sprayed against the belt after the main material load has been shed will act as a release agent, diluting the material and preventing adhesion to most surfaces. It will also act as a coolant to the secondary cleaner blades to prevent the "baking on" of the material. As a bonus, the life of cleaning blades can be extended by the lubricating action of the water.

In the paper "Performance Guaranteed Belt Cleaning," D. J. Bennett reported that "Belt cleaning systems incorporating water improve results by a factor of three over their respective non-water spray systems."

The results gained by the correct application of water in belt cleaning systems more than justifies its consideration in any suitable materials handling plant.

Belt Washing Stations

Two styles of cleaning systems use water applied to the belt through a spray wash enclosure or "washbox" to assist in the cleaning process. **(Figure 13.32)** The first directs a low-pressure 5 to 10 psi (.035 to 0.7 bar) mist of water at the belt to assist in the scraping process. The second washing system uses a higher pressure 40 to 50 psi (2.7 to 3.4 bar) spray of water to remove the carryback from the belt. After the nozzles wet the belt, one or more cleaner assemblies with urethane or metal blades remove the material and water from the belt. Either style of "washbox" can provide effective cleaning, but they both pose the problem of disposing of the slurry consisting of removed material and water. In some cases, the water can be filtered and reused; in other cases, both water and material can be returned to the conveyed material load. The low-pressure system uses less water and, as a result, reduces the sludge-handling problem. Both systems must be carefully engineered to avoid problems in cold weather operations.

Where the application of water will result in increased moisture in the material, awareness of the volume added is required. However, generally the total added moisture is not substantial, especially when it is understood that much of the water will not actually stay in the material. Mist losses at the application point and evaporation on the belt and stockpile will probably reduce the added moisture by half.

The use of water is an efficient way of removing remaining material from the belt surface. However, it does not suffice to put or spray water onto the belt; it has to be ensured that the water/dirt slurry created is completely removed from the belt and properly disposed of or settled out.

In a post-discharge belt washing station, the amount of added water is not so critical since it is not actually added to the process or the main material body. Rather, the controlling factor will be sizing of sumps and pumps to dispose of or recycle the slurry.

Belting and Belt Cleaners

Belt cleaning can be made difficult by patterns on the belt surface as typically seen in PVC belting. It is also difficult to clean a belt that has cracked or is "alligatoring" due to exposure to the elements or chemical attack. In both cases, the only method to remove the material is washing the belt.

Some belt manufacturers continue the ill-advised embossing of identification numbers and corporate logos into the covers of the belt. It is easy to appreciate the marketing value of this practice. However, it is just as easy to recognize the difficulties that these emblems in the face of the belt can create in effectively cleaning and sealing the conveyor system.

Steel cable belts often show on their surface the pattern of the cables hidden inside the rubber. This gives the effect of hard and soft streaks in the belt. To remove the "streaks," cleaning pressure is increased beyond what is necessary and cover wear is accelerated.

All methods of belt cleaning and all blade materials wear the belt's top cover. At least one manufacturer of belting incorporates a factor for this wear into the design of the top cover. It is generally accepted that it is better to wear the belt slowly by cleaning than rapidly by dragging it through dirt and across non-rotating or damaged idlers.

In reality, a well-designed belt cleaning system has much less negative effect on cover life than does the loading of the material on the belt. Top cover selection should be governed by the considerations of material loading, rather than worries over belt cleaning.

Trouble Spots on the Belt

Often, a conveyor belt in generally "good condition" and relatively easy to clean will suffer from a longitudinal groove type of damage that renders conventional cleaning methods ineffective. This condition is particularly annoying because the belt life is not affected, but the belt will be a problem to clean for its remaining life. These grooves can be cleaned effectively by a "spot cleaner," with blades located only in the path of the grooves, while allowing for the normal, slight transverse movement of the belt. This "spot cleaner" would be located after the "whole belt" cleaning system.

Belt Cleaners and Horsepower Requirements

Applying one or more cleaners against the belt increases the drag against the belt and so raises the conveyor's power consumption. As reported by the Bureau of Mines, over-tensioning the cleaners is of doubtful cleaning value and will dramatically increase horsepower requirements. However, the consequences of not installing and properly maintaining effective cleaners can prove a more serious drain of conveyor drive power through the added friction caused by idlers with material accumulations or seized bearings.

A study by R. Todd Swinderman published in *Bulk Solids Handling* has examined how many horsepower the application of a belt cleaner consumes from a conveyor's overall drive power.

This paper calculated the power requirements for five types of cleaners, using the following formula:

Force (lbs) = Blade Coverage (in) x Load (lbs/in)

$$\text{Power (hp)} = \frac{\text{Force (lbs) x Coefficient of Friction x Belt Speed (fpm)}}{33000}$$

In metric units, the equation is:

Force (N) = Blade Coverage (mm) x Load (N/mm)

$$\text{Power (kW)} = \frac{\text{Force (N) x Coefficient of Friction x Belt Speed (m/s)}}{1000}$$

The coefficients of friction used in these calculations are derived from information supplied by belt manufacturers and are based on tests on new, dry, Grade 2 rubber belt samples against urethane and steel blades. The static coefficient of friction was used as it

Blade Type	Belt Speed, FPM (m/sec)			
	100(.5)	*400(2)*	*700(3.5)*	*1000(5)*
	hp(kW)	hp(kW)	hp(kW)	hp(kW)
Metal-Bladed Secondary Cleaner *	0.2 (0.14)	1.0 (0.75)	1.7 (1.27)	2.4 (1.79)
Urethane-Bladed Pre-Cleaner**	0.2 (0.14)	0.7 (0.52)	1.3 (0.97)	1.8 (1.34)
Metal-Bladed Secondary Cleaner**	0.3 (0.22)	1.2 (0.89)	2.1 (1.57)	3.0 (2.24)
Urethane-Bladed Secondary Cleaner **	0.5 (0.37)	2.0 (1.49)	3.6 (2.68)	5.1 (3.80)

* tensioned with "optimum pressure" specified by U.S. Bureau Of Mines in research cited above.
**tensioned with air tensioner supplied by Martin Engineering to recommended cleaning pressure.

Figure 13.33

Power consumption added to conveyor drive requirement by various types of belt cleaners.

gives an estimate of the start-up horsepower. The running horsepower can be estimated as approximately two-thirds of the start-up horsepower.

Note: Power requirement is calculated for the belt width actually contacted by the cleaner. In most cases, the cleaning blade does not contact the full width of the conveyor belt.

As a sample of his results, let us consider a 36-inch- (900 mm-) wide belt moving at speeds of 100, 400, 700, and 1000 feet per minute (0.5, 2, 3.5, and 5 meters per second). The power consumption added to the conveyor's drive by the tensioning of various types of belt cleaners is presented in **Figure 13.33**. Blade coverage of the cleaners against the belt is 30 inches (762 mm).

In Swinderman's final example, he considers a typical conveyor application: a 48-inch- (1219 mm-) wide belt, operating at 600 feet per minute (3 m/sec), conveying 1500 tons (1365 metric tons) per hour of coal for a distance of 300 feet (91 meters) at an incline of 14 degrees. The weight of the belt is specified as 15 lbs/ft (22.3 kg/m), and idlers are spaced every two feet (600 mm). As calculated by one commercially available conveyor engineering computer software program, this conveyor would require a total drive power of 138 hp (103 kW).

If an extreme amount of carryback–say 0.25 lb./square foot (1.2 kg/square meter) were present on the belt, this would amount to 12 additional tons (10.9 metric tons) per hour of load, and in itself would require very little additional power to carry.

However, further investigation reveals conveyor problems do not arise from the horsepower consumed

$$HP = \frac{\text{Weight} \times f \times \text{Speed}}{33,000}$$

Coefficient of friction (f) = 1.0 – 0.1 = 0.9 increase (assuming normal roller = 0.1)

$$HP = \frac{(83.33 \text{lb/ft} + 15 \text{ lb/ft}) \times 0.9 \times 600}{33,000} = 1.6 \text{HP} = 1.19312 \text{kW}$$

Figure 13.34

Calculation for frozen impact idler set.

$$HP = \frac{\text{Weight} \times f \times \text{Speed}}{33,000}$$

Coefficient of friction (f) = 0.2 – 0.1 = 0.1 increase (assuming rubber on polished steel =0.2 and that a normal roller = 0.1)

$$HP = \frac{(83.33 \text{lb/ft} + 15 \text{lb/ft}) \times 0.1 \times 600 \text{ fpm}}{33,000} = 0.36 \text{HP} = 0.268452 \text{kW}$$

Figure 13.35

Calculation for frozen steel roller set.

by the carryback, but rather the impact on the conveyor hardware of this carryback as it is released into the environment.

For example, Swinderman calculates that a single frozen impact idler set would require approximately 1.6 horsepower (1.2 kW) additional. **(Figure 13.34)** One seized steel idler set can demand as much as 0.36

horsepower (0.27 kW) additional. **(Figure 13.35)** The Swinderman study also noted that a one inch (25 mm) layer of carryback on a single return roller can add as much as 0.43 additional HP (0.32 kW) to the conveyor's drive requirements.

These additional power requirements for the problems arising from fugitive material should be compared with the horsepower requirements of a typical dual cleaning system. Continuing Swinderman's example above–a 48-inch (1219 mm) belt traveling 600 FPM (3 m/sec) and now incorporating a dual cleaning system–the horsepower requirement would be 1.7 hp (1.3 kW) for the pre-cleaner and 2.8 hp (2.1 kW) for the secondary cleaner.

The combined total of 4.5 additional horsepower (3.4 kW) required for the use of an effective multiple cleaner system represents an increase of only three percent over the 138 hp required for the conveyor without any cleaners. The "conveyor power penalty" applied by the belt cleaning system is only a little more than the power consumed at the rate of 0.36 horsepower (0.27 kW) for a single seized idler set or 0.43 horsepower (0.32 kW) required by one inch (25 mm) of accumulation of material on one return roller. It won't take many frozen idlers before the power consumption is the same. And of course, fugitive material will not stop with affecting a single idler. The material buildup can quickly interfere with all the rolling equipment on the conveyor.

These findings indicate the drive power requirements for cleaning a conveyor are significant but less than the consequences of not installing and maintaining an effective engineered cleaning system. In other words, it will cost a little power to apply a belt cleaning system. However, if a cleaning system is not installed, the consequences to power consumption will soon be similar, and over time will get worse. It is a case of "pay me (a little) now, or pay me (much more) later."

Belt Cleaners and Belt Life

Conveyor belting is a major operating expense and any effort to improve the service life of the belt will have a significant impact on plant profitability. The question of how much the use of an engineered belt cleaning system against the moving belt will shorten the life expectancy of a belt is worthy of some consideration.

The mechanisms of wear depend a great deal on the amount of heat generated in the blade and top cover. Field observations indicate–particularly for elastomeric blade materials–the highest wear rate to both blade and belt occurs when there is no material on the belt.

"Belt Cleaners and Top Cover Wear," a research study by R. Todd Swinderman and Douglas Lindstrom, has examined the issue of whether belt cleaners adversely affect the life of the belting. The results of this study showed a belt cleaner can induce wear of the belt, but the rate of wear was still less than allowing the belt to run through abrasive carryback without benefit of cleaners.

These findings also support the field observation that the wear process is roughly a linear relationship, depending on the pressure used and the amount of mass available for wear in both the belt and the blade. This testing also supported the Bureau of Mines' finding that there appears to be an optimum pressure for minimizing blade and top cover wear.

Similar results were reported in a study that compared belt life and belt failures in facilities using engineered cleaning and sealing systems as opposed to facilities that did not use these systems. Performed in India, this survey reviewed over 300,000 hours of operation on 213 belts in facilities handling lignite, limestone, and iron ore. The study showed that facilities using the engineered systems had belts that lasted an average of 150% longer (and required 50% of the cleaning labor) than the facilities not equipped with engineered cleaning and sealing systems. This survey of operating facilities echoes the laboratory research indicating, while belt cleaning systems do introduce some wear to the belt cover, the end result is the cleaner the belt, the longer it will last.

Cleaner Installation

Regardless of the make of a belt cleaner the critical factor in cleaner installation is that the center of the cleaner mainframe must be installed at the correct distance from the face of the pulley. Maintaining the proper dimension places the blades at the correct angle of attack against the belt for best cleaning, proper blade wear, and longest life.

This distance will be different from cleaner style to cleaner style, because the length and curve of blade design varies. In some cleaners, this dimension also varies in accordance with the diameter of the head pulley.

Considerations in affecting the installation position of a belt cleaner include:
- Cleaner Design.
- Tensioner or Mount Requirements.
- Bolting or Welding the Cleaner in Place.
- Installation on Chute Wall or Hung from Stringer.
- Position of Conveyor Structural Beams, Bearings, and Drives.

Be sure to consult the manufacturer's installation or operator's manual for all belt cleaner installation instructions.

Pre-Cleaner "Heeling"

Many problems can occur when a pre-cleaner blade is mounted too close to a belt. If the pre-cleaner blade has a flat or curved contact surface, a cleaner mounted too close will cause the heel of the blade to contact the belt first. This "heeling" creates a gap along the leading edge of the blade. **(Figure 13.36)** Material will collect in this gap to the extent that the accumulation will force the blade away from the belt. Once the blade is forced away from the belt, large amounts of material will pass between the belt and the blade. This greatly decreases cleaning efficiency. The passage of this amount of material will also cause the blade to wear much faster. It is very important that the pre-cleaner be installed to the manufacturer specifications. Failure to do so will result in ineffective cleaning and greatly reduced wear life.

Troubleshooting Cleaner Installation

If the blades do not show excessive wear and the tensioner is set correctly, but the cleaner is not cleaning sufficiently, other problems may be present. These problems might include:
- The mainframe is not parallel with the pulley.
- The cleaner is not installed the proper distance from the belt surface.
- The cleaner is changing the belt line.
- The blades are not centered on the belt.

Any of these factors will impair a cleaner's ability to remove carryback. Review the cleaner manual to determine appropriate corrective actions.

The Importance of Maintenance

Even the best-designed and most efficient belt cleaning systems require maintenance and adjustment on a more-or-less regular basis. Cleaners will need to be retensioned against the belt to accommodate blade wear. Material accumulations should be knocked off and worn blades replaced.

Often, you get out of a belt cleaning system what you put into it. If a cleaner is installed and never looked at again, its performance will deteriorate over time. If inspection, adjustment, and maintenance are provided on a timely basis, cleaning will continue to match original standards. Proper maintenance of belt cleaning systems will reduce wear on the belt and cleaner blades, prevent damage, and ensure efficient cleaning action. Lack of maintenance on belt cleaner systems not only produces a failure to clean effectively but adds considerable risk to the conveyor. Consequently, it is important that both the conveyor and the cleaning system be engineered to facilitate this essential maintenance.

It is–or at least should be–one of the commandments of belt cleaner design that cleaners be easy to service. Manufacturers must design their equipment to simplify maintenance activities. Maintenance requirements and procedures should be reviewed during the selection process for a cleaning system. Advance planning for cleaner service will allow maintenance activities to be performed expeditiously with minimum discomfort and downtime. This will translate into improved belt cleaning, as it is likely that service chores that are simple and "worker friendly" will actually be performed on a consistent basis. The final result will be improved conveyor efficiency.

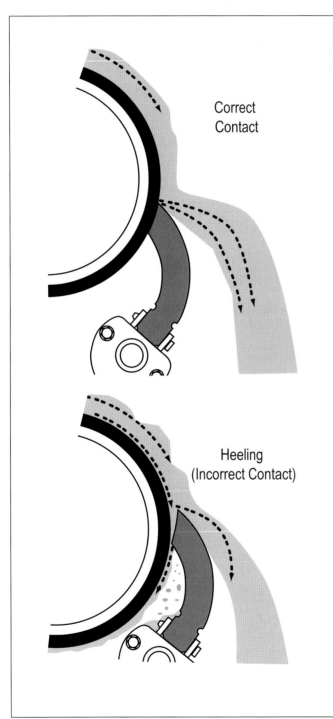

Figure 13.36
When a pre-cleaner is installed too close, the resulting "heeling" leads to ineffective cleaning and reduced blade life.

It is readily apparent that good ergonomic design is important to allow an operator or millwright to comfortably access the cleaners. This will allow personnel to make inspections and provide appropriate service as needed, perhaps even several times a day on the occasions when the material is in its "worst case" condition. The conveyor itself should be designed to provide adequate access with ample clearances and work spaces. Access windows with easy-to-operate doors should be installed on both sides of the pulley and in line with the axis of the belt cleaners.

Elements can be incorporated into a conveyor belt cleaning system to enhance maintenance procedures. These would include easy-opening access/inspection doors, cleaning elements that slide in and out for service without needing the removal of the mainframe from cleaning position, and blades that resist corrosion and abuse, that can be installed using only simple hand tools. These considerations mean that any operator or inspector can quickly perform the necessary service without waiting for a maintenance crew with power tools to make time to perform the work.

Mandrel mounting systems that allow slide-in/slide-out positioning of a cleaner assembly offer an opportunity for faster service. **(Figure 13.37)** Also available are cleaning systems mounted on mainframes that incorporate telescoping sections, so the whole assembly goes into position. **(Figure 13.38)** Some facilities have made arrangements for this service to be performed while the belt is running, given regulatory and safety committee approvals and proper personnel training.

Belt Cleaner Maintenance

Even the best-designed and most efficient belt cleaning systems require maintenance and adjustment on a regular basis. In many cases, a plant will get out of a cleaner what they put into it. In that vein, it may be best to have a plant's belt cleaning systems selected by the plant personnel, as they will then have made a commitment to it and will work on its behalf. The most effective belt cleaners can be those designed or selected by the operators, because then they will have a stake in the cleaners' success.

Specific maintenance instructions are provided in the manual for each cleaner and tensioner. Each belt cleaner maintenance procedure should include the following:

Clean Off Accumulated Material

With the belt stopped, clean off any material that has gathered between the cleaner blades and the belt or built-up on the arms of secondary cleaners. Often, a rotation of the cleaner away from the belt followed by a few sharp raps of the blades back into the stopped belt will dislodge this material. In other conditions, a quick rinse with a water hose or high-pressure spray will remove the buildup and allow inspection of the blade.

Check Cleaner Performance

Is the belt clean? Carryback remaining on the belt could indicate worn out blades or improper tension.

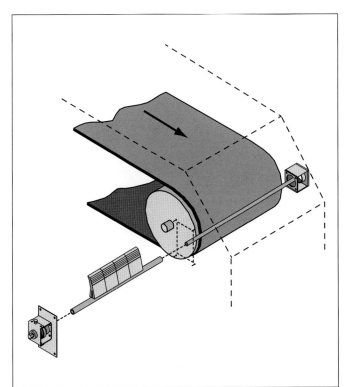

Figure 13.37
A mandrel-style belt cleaner mount allows the cleaner to slide out on a stationary beam.

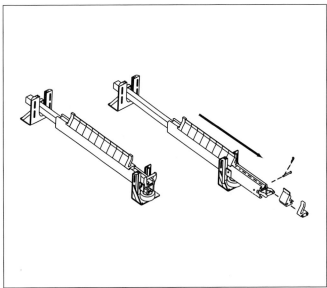

Figure 13.38
A telescoping belt cleaner assembly offers improved service efficiency.

Check Blade Wear

Are the blades worn? Some brands of blades incorporate a visible wear line; for others you will need to check the manual for the limits of safe and effective wear.

Check Tensioner Adjustment

The most critical element in maintaining cleaner performance is keeping the blades tensioned against the belt. As blades wear, the tensioner may need to be adjusted to accommodate the blade's shorter length. Each tensioner's manual should provide specific instructions for re-tensioning the belt cleaners.

Always lock out and tag out power to the conveyor and any accessory conveyor equipment before beginning any work in the area of the conveyor.

Improving Cleaner Performance

There are a number of methods that can be recommended to upgrade the performance of a plant's belt cleaning systems.

- Belt repair kits are available to correct problems with cuts, gouges and cracks in the belt surface.
- Mechanical splices can be dressed or recessed to prevent damage to cleaner and improve cleaning performance.

A general program for improving belt cleaner performance might include:

1. *Follow the Instructions*

   Make sure systems are installed and maintained to the manufacturer's recommendations. Observing the maintenance interval is of particular benefit.

2. *Standardize and Systemize*

   You can standardize the cleaners on all conveyors in your plant and/or all plants in your company using identical installations or a system that incorporates options to fit all circumstances. In addition, you can systemize cleaner maintenance, either by managing it inside the plant or by outsourcing it to a specialized contractor–to improve execution and accountability.

3. *Raise the Bar*

   Aim for continuous upgrading of performance. Acquire additional understanding of cleaning process and cleaning performance through measures such as a blade-life testing program to improve blade selection or by considering the installation of additional cleaners on problem conveyors.

It is now possible to analyze cleaning systems to identify the blade-to-belt pressure that optimizes both cleaning efficiency and blade life. To allow this testing, some belt cleaners are marked with wear percentages molded into the blades. Tensioning systems have been developed that provide continuous constant spring pressure. The test consists of setting the cleaning pressure at a given level and recording the length of time it takes for the blade to reach the 25% wear mark. The pressure is then adjusted, and the cleaner used until it wears to the next 25% wear line. The cleaning efficiency should also be studied, using the quantitative methods discussed earlier in this chapter or with a visual qualitative measurement. The results can be measured against time; that is, in days of operation or against the tonnage of material carried during the test. In this way, the plant can determine what pressure provides longest life while maintaining cleaning efficiency.

The results will vary from application to application, and even from conveyor to conveyor within the same plant. One German lignite mine found that a higher cleaning pressure resulted in a longer blade life; a similar operation reported that lower pressures resulted in longer blade life while maintaining acceptable levels of cleaning. This type of study provides a method to establish the optimum efficiency and life and so improve cleaner performance.

Handling Material Cleaned from the Belt

The fact that carryback clings to the belt past the discharge point indicates that it has different characteristics than the rest of the load. The particles are finer, with a higher moisture content and different flow qualities than the main cargo. It is not unusual for carryback material to adhere to the surface of a low-friction liner. Even after its removal from the belt, carryback material presents problems in capture, handling, and disposal.

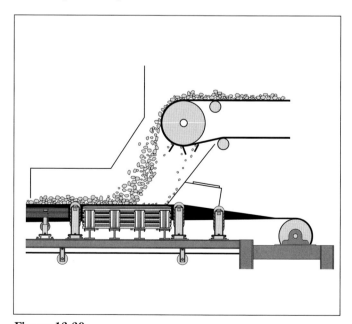

Figure 13.39

Dribble chutes return the material cleaned from the belt to the main cargo.

Figure 13.40

Material can build up in the dribble chute and impair the operation of the cleaner or the belt.

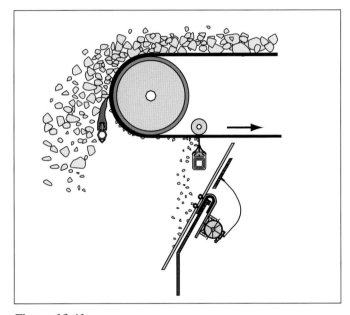

Figure 13.41

A vibrating dribble chute uses an electric vibrator to activate a plastic "false wall" inside the chute.

Figure 13.42

The rubber-lined mounting bracket transfers vibration to the plastic liner while protecting the steel chute.

As noted above, it is usually best to locate the belt cleaner(s) on the conveyor return as far forward—as close to the discharge point—as possible, without being in the trajectory of the discharging material. This gives the belt most, if not all, of its cleaning before fugitive material can be released into the plant. In addition, cleaning on the head pulley minimizes the difficulty in handling the material cleaned off the belt.

The collection and return to the main material body of the carryback removed by conveyor belt cleaning devices can present a serious complicating factor in design of discharge chutes. Ideally, the conveyor's main discharge chute is sufficiently large, so the material cleaned from the belt can fall through the same chute, where it is reunited with the main stream of material. If the main chute is not large enough, an extra fines chute or dribble chute should be added under the secondary cleaners. This auxiliary chute must be large enough and designed with a steep wall angle to prevent the encapsulation of the cleaner in these sticky materials.

Dribble Chutes

On conveyors where the cleaning systems are positioned so that the material they remove from the belt does not freely return to the main material trajectory, a dribble chute is required. This is a separate chute that deflects the cleanings back into the main material flow. **(Figure 13.39)** A dribble chute under a belt feeding into a transfer point should have a slope that is based not upon the slope of the other chutework but rather on testing of the carryback material to determine its cohesiveness. Whenever possible, it is advisable to install vertical dribble chutes, as they are the best at minimizing problems from material buildup.

To ensure continuing performance, the buildup of fines on the cleaner or dribble chute must be prevented. **(Figure 13.40)** A cleaner that is encapsulated with a sticky or cured accumulation of material cannot function. Similarly, a chute that is clogged with material will fail by allowing the material to back up and eventually bury the cleaner. Often, a belt cleaner is judged to be inefficient, when in reality, scraped-off carryback fines have built up in the dribble chute and submerged the cleaning device. Consequently, dribble chutes and other fines reclamation systems must be carefully designed and constructed.

It may be useful to incorporate mechanical aids to help in moving the carryback material away from the cleaners. Options to assist the movement of material would include low friction chute liners, or the installation of vibrators on cleaner mainframes, or on the wall(s) of the chute.

One way to solve the problem of dribble chute buildup is to create a dynamic sub-floor inside the chute. This can be accomplished by mounting a sheet of smooth, low-friction, abrasion-resistant plastic such as UHMW Polyethylene, so the plastic is parallel to the chute floor but free to move. **(Figure 13.41)** A U-shaped piece of metal is used to support the plastic sub-floor at one end; the other end is left unsupported. By mounting a vibrator to the free leg of the bracket, its vibrating action is transmitted through the leg to the plastic subfloor, providing a dynamic action to prevent material buildup. **(Figure 13.42)** As this installation is isolated from the chutework by a rubber lining inside the mounting bracket, there is no force applied to the structure to cause metal fatigue.

An alternative system would include a flexible curtain or sheet of rubber used as a chute liner, which is periodically "kicked" with the discharge from an air cannon. This flexes the liner, causing any adhered material to drop away. *See Chapter 5: Loading Chutes.*

Access into the dribble chute should be provided to clear buildups and allow a periodic wash down to prevent blockages.

Scavenger Conveyors

When the belt cleaning devices are mounted along the conveyor away from the discharge point, it may be beneficial to provide an auxiliary conveyor to return the cleanings to the main material flow. Small screw conveyors, scraper-chain conveyors, **(Figure 13.43)** or vibratory conveyors **(Figure 13.44)** are commonly used as scavenger systems.

Pulley Wipers

It is possible for materials that will stick to the belt to transfer and adhere to any snub or bend pulleys that contact the dirty side of the belt. Therefore, it may be necessary to install belt cleaners as pulley-cleaning devices.

To clean adhering material from a pulley, a scraper is typically mounted on the ascending side of the pulley, about 45° below the horizontal centerline. **(Figure 13.45)** This position allows material scraped from the pulley to fall freely. Care must be taken to provide a dribble chute or an accessible area from which this material can be cleaned when it accumulates.

Belt Turnovers

To eliminate the problems caused by a dirty belt contacting the return idlers, the belt can be twisted 180° after it passes the discharge point. This brings the clean surface of the belt into contact with the return idlers. The belt must be turned back again 180° before it enters the tail section to return the carrying side of the belt to the up position at the loading point. The

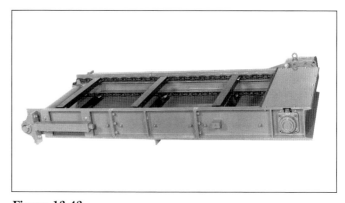

Figure 13.43

A drag conveyor can be used as a scavenger conveyor underneath a belt cleaning system.

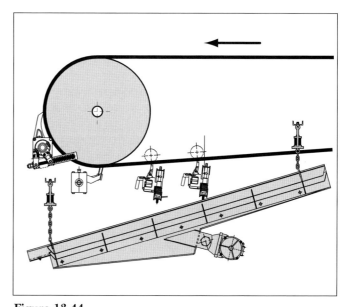

Figure 13.44

Vibrating conveyors can be installed as scavenger conveyors to return material to the transfer point.

Figure 13.45

This belt cleaner is installed to remove material from a pulley.

distance required to accomplish the 180° turnover of the belt is approximately 12 times the belt width at each end of the belt.

Turnovers are generally installed after the conveyor's take-up system or snub-pulley system, and they generally leave plenty of opportunity for carryback to fall from the belt into conveyor components before the turnover system inverts the belt.

The act of twisting the belt can still release fugitive material. This motion can cause adhering material to "pop off" the belt, like twisting the ends of a wrapper can cause a piece of hard candy to jump out of the wrapper. Because this release takes place at the point where the belt is turned over and, therefore, the point where idler alignment and cleanliness are most crucial, it is particularly problematic.

To minimize the carryback that could be released when the belt is twisted, an effective belt cleaning system must be installed at the conveyor's discharge.

chapter 14

Tail Protection Plows

"In a conveyor system where stability is a key to control of fugitive material, any damage to the belt or pulley can adversely affect the performance."

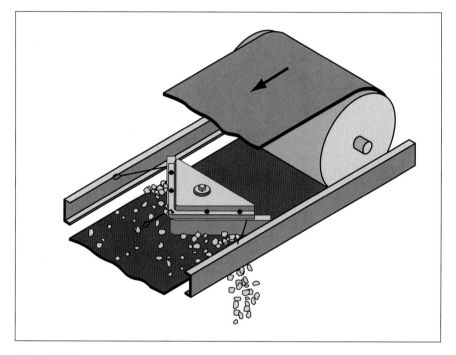

Figure 14.1

Tail protection plows prevent the entrapment of material lumps between the belt and the pulley.

As the conveyor belt returns from its discharge point at the head pulley to the loading zone, it passes around the tail pulley. Occasionally, as it moves back, the inner side of the returning belt will carry a lump of spilled material into the tail pulley. The entrapment of this lump between belt and pulley can do dramatic damage, puncturing the cover of the belt or damaging the lagging or pulley face. Even small particles and fines can wear and grind away on the less durable, more easily damaged inside surface of the belt. In addition, material can build up on pulleys to cause belt slippage, mistracking, and lead to the risk of fire.

In a conveyor system where stability is a key to control of fugitive material, any damage to the belt or pulley can adversely affect the performance.

The basic protection against this material impingement is proper control of loading. The correct loading angle and drop height, the proper match of loading speed to belt speed, and the proper loading direction in relation to belt travel all are factors that help settle the load, reduce

Figure 14.2
A new diagonal plow design features a curved blade to discourage material accumulation.

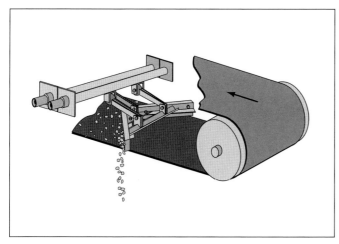

Figure 14.3
V-Plows are installed with the point of the "V" pointing away from the tail pulley.

Figure 14.4
Plows should provide firm but flexible cleaning pressure, allowing them to float with fluctuations in the belt's line of travel.

agitation, and minimize material spillage. Good belt training is fundamental, because misaligned belts also produce spillage that can put material on the return side of the belt.

But even in the most ideal installations, some material may spill off the load and onto the inside of the belt. To protect against this possibility, tail protection plows or return belt scrapers are installed.

Tail Protection Plows

Positioned on top of the return side of the belt near its entrance to the tail pulley, these plows deflect lumps of material off the belt. **(Figure 14.1)** Plows prevent large objects and tramp iron from damaging the belt and tail pulley. They remove these fugitive materials with a simple, low-pressure scraping that directs the material off the belt. Installed between the tail pulley and the first return idlers, tail plows represent a form of low cost "insurance" when weighed against the out-of-pocket costs of conveyor maintenance, damage, and possible premature replacement of the belt, as well as the costs of unscheduled downtime. **(Figure 14.2)**

Depending on conveyor configuration and load, it may be desirable to install additional plows to protect other pulleys on the conveyor.

Considerations in Plow Design

When specifying a tail plow, there are a number of factors that should be considered. A plow should:

1. *Provide firm but flexible pressure*
   To allow the plow to float across the belt surface, removing materials–both fines and larger pieces–effectively yet adjusting automatically to accommodate wear in the plow blade and fluctuations in the belt path. **(Figure 14.3)**
2. *Be securely mounted*
   To minimize the risk of it breaking away from its mounting system, endangering the conveyor components it was installed to protect. (It should also be fitted with a safety cable to protect the belt and pulley should unexpected failure of the plow mounting occur).
3. *Be designed for ease of installation*
   To minimize downtime. (For example, the supporting arms should fit within the existing conveyor structure without modification).
4. *Allow easy replacement of blade material*
   To provide a long service life and fast maintenance.
5. *Be readily accessible*
   To allow observation and maintenance.

Plow locations should be carefully sited, so that the material removed by the plow does not create a hazard as it falls or where it accumulates.

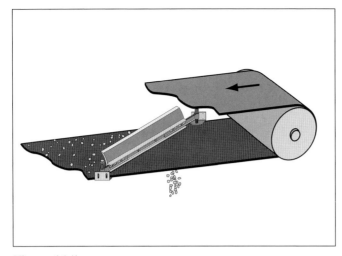

Figure 14.5

A recent development is a plastic V-plow with a hollow tank to allow adjustment of its weight.

Figure 14.6

A V-plow is typically installed on single-direction conveyors near the point where the belt enters the tail pulley.

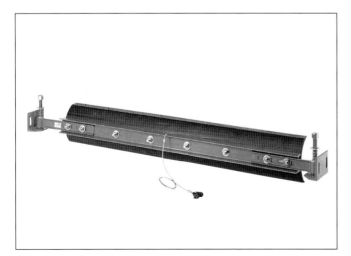

Figure 14.7

A diagonal plow can be used on a belt that reverses or has significant rollback.

To prevent the large lumps from "jumping" over the plow (and then becoming hung up in the plow suspension or conveyor structure, or becoming trapped between belt and pulley), plows should stand as tall as the largest lump conveyed with a minimum height of four inches (100 mm). It is advantageous to cover the "inner space" of the plow, to prevent material from becoming caught in the plow itself.

V-Plows and Diagonal Plows

Typically, this scraper is a diagonal plow or V-plow with a steel frame, using a rubber, urethane or plastic blade to push any fugitive material off the belt. In most cases, the plow rides on the belt, tensioned against the belt by its own weight, and floats up and down with any fluctuations in belt travel. The plow should be secured with a safety cable to prevent it being carried into the pulley should it become dismounted.

The plow should be suspended from a point above and in front of its point. This is to ensure that the impact of lumps does not force the plow downward into the belt. To prevent the plow from somersaulting or inverting, a holddown on the tail of the plow may be required.

On belt conveyors that travel only in one direction, the return belt cleaner is usually a V-plow. **(Figure 14.4)** The point of the "V" is aimed away from the tail pulley, so any loose material carried on the belt's inside surface is deflected off the conveyor by the wings of the plow.

If the conveyor, through either reversing operation or accidental rollback, has two directions of movement, the return belt scraper should be a diagonal plow that provides protection in both directions. **(Figure 14.5)** Diagonal plows are installed across the belt at an angle of 45° to the direction of travel. If the belt truly runs in two directions, so either pulley can serve as a tail pulley, diagonal plows should be installed at both ends of the conveyor.

A diagonal plow can be installed on a one-direction conveyor, if there is a reason to collect or discharge the material on a specific side of the belt. In this case, the diagonal plow should be installed so its discharge end (the side of the belt where the cleaned material should be collected) is closest to the pulley.

Advantages of Engineered Plows

While plows themselves are fairly simple devices, there are some twists that demonstrate the advantages of using engineered systems rather than homemade scrapers. The engineered systems can provide a long-term solution that is economical in performance, service-life, and maintenance expense, over the false economies of the homemade unit. Now, with innovations in design and construction, engineered systems are available to provide the benefits of an engineered system while minimizing the initial investment.

One innovation is a unit incorporating a hollow compartment or tank that is filled with ballast, such as gravel or sand. This allows the unit's weight to be varied to create optimum stability and cleaning performance. **(Figure 14.6)** This plow uses an all-plastic construction to minimize costs.

Another type of diagonal plow uses a blade with a curved face, rather than the typical straight blade installed perpendicular to the belt. **(Figure 14.7)** This curve ensures that even the most cohesive material cannot build up on the face and climb the blade. Rather, material will drop off the curved surface back onto the belt where the plows diagonal installation will direct the material off the belt. **(Figure 14.8)**

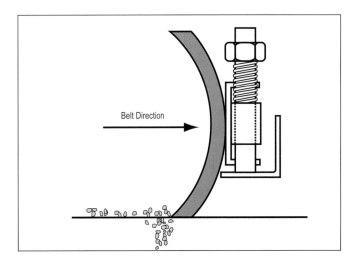

Figure 14.8

The curved plow face prevents cohesive material from climbing the blade.

Chapter 15

Keeping the Belt on Track

"To protect against damage and provide a proper, long-term return on this investment, it is essential to maintain proper belt tracking."

Figure 15.1

Belt mistracking is one of the easiest problems to spot, but one of the more complicated to remedy.

In an ideal world, a sound belt on a well-engineered and maintained conveyor structure should not wander. However, mistracking belts are an everyday fact of life in many bulk material handling operations, and can pose very expensive problems. **(Figure 15.1)** A conveyor that does not run in alignment can cause material spillage, component failure, and suffer costly damage to its belt and structures. Allowing a belt to run to one side of the structure **(Figure 15.2)** can greatly reduce its service life, as it becomes stretched or folds over on itself. Expensive belts can run against steel chutes and structural members to the point where the belts **(Figure 15.3)** and even the steel structures **(Figure 15.4)** are damaged, often beyond repair.

Belts that are hundreds of meters long and cost several hundred dollars per meter can easily represent a total replacement cost of one million dollars. To protect against damage and provide a proper, long-term return on this investment, it is essential to maintain proper belt tracking.

In many ways, proper belt tracking is a precursor to and a fundamental requirement for resolving many of the fugitive material problems discussed in this volume. Belt tracking must be controlled before spillage can be eliminated; if the belt wanders back and forth through the loading zone, material is more readily released under the skirtboard seal on either (or both) sides. Just as the belt path must be stabilized horizontally through proper belt support and the elimination of wing pulleys, the path must be controlled in the horizontal direction through the elimination of belt wander.

An Avoidable Problem

Mistracking is a system problem. A wandering belt should serve as a flag, signaling there is something more seriously wrong with the conveyor system.

As Clar Cukor noted in the Georgia Duck's monograph *Tracking*,

> [A conveyor belt] is flexible and if designed, manufactured, slit and cut properly, will "go where directed" by the conveyor system as designed and built. The problem of tracking should be approached from a systems point of view. The belt may well be at fault–however, it is more likely merely reacting to a structural defect or maladjustment in the system. The conveyor belt serves as an indicator and should be so regarded.

Wandering belts can be caused by a number of problems. Factors contributing to belt wander include misalignment of conveyor components, off-center loading of cargo, accumulation of fugitive material on rolling components, poor belt splices, structural damage caused by inattentive heavy equipment operators, ground subsidence, and many others. And of course, these problems may be inflicted on the system in any combination, greatly complicating the process of correction.

Causes of Belt Wander

In many cases, the cause of the mistracking can be determined from the form the mistracking takes. When all portions of a belt run off at a certain part of the conveyor length, the cause is probably in the alignment or leveling of the conveyor structure, idlers, or pulleys in that area. If one or more sections of the belt mistrack at

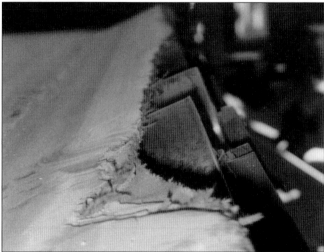

Figure 15.3
This return idler bracket has nearly been cut through by the edge of the mistracking belt.

Figure 15.2
Belts that run to one side of the structure risk expensive edge damage.

Figure 15.4
Edge damage resulting from the belt mistracking into the conveyor structure can shorten the belt's service life.

all points along the conveyor, the cause is more likely in the belt construction, in the splice(s), or in the loading of the belt.

Sometimes, a combination of problems will produce belt wander where the cause is not clear-cut. However, if a sufficient number of belt revolutions are observed, the running pattern generally will become clear, and the cause of mistracking disclosed. The usual cases when a pattern does not emerge are those of erratic running from an unloaded belt that does not trough well or a belt that is unevenly loaded.

The most common causes of mistracking can be split into three groups:

1. Faults with the belt:
- the splice is not square to the belt.
- the belt is bowed or cupped.
- the belt has a camber.
- poor installation of vulcanized or mechanical splice.
- defects or damage in the carcass (plys or cords) of the belt.
- top cover damage such as tears or holes.
- different types or thicknesses of belt have been spliced together.
- the belt is poorly matched to the conveyor structure (i.e., the belt's troughability is unsuitable).
- degradation of the belt from exposure to the elements or chemicals.

2. Faults with the conveyor structure:
- the structure was not accurately aligned during its construction.
- idlers and pulleys are not aligned in all three axes.
- material buildup alters the profile of idlers or pulleys.
- idler rolls have seized or been removed.
- the conveyor is subjected to lateral winds.
- the conveyor is subjected to rain or sun on one side or to other adverse environmental conditions.
- the gravity take-up is misaligned.
- the structure has been damaged from collisions with mobile equipment (front-end loaders, haul trucks, etc.).
- the structure has settled on one side through ground subsidence.
- the structure has warped due to a conveyor fire.
- storm damage to structure.

3. Faults with material loading:
- the load is not centered.
- the load is segregated, with larger lumps on one side of the belt.
- intermittent loading on a belt that is tracked for a constant load.

Here are some typical causes of mistracking:

Belt Wander Due To Bad Splices

Improper belt splicing is a significant cause of mistracking. If the belt is not spliced squarely, the belt will wander back and forth on the conveyor structure. This can usually be seen at the tail pulley. The belt will wander the same amount each time the splice reaches the tail pulley, only to return to its original position after passage of the splice. If the splice is bad enough, it can negate all alignment efforts. The solution is to resplice the belt squarely.

Placing a carpenter's square against one side of the belt may not give a true evaluation of the space, as belt sides might not be truly parallel due to faults in its manufacture or slitting. The proper way to splice is to establish a centerline at each belt end, match the centerlines so they form a continuous straight line, and make all square measurements based on that now-shared procedure.

Mistracking Due To Environmental Conditions

Strong winds on one side of the conveyor can provide enough force to move the belt off line. The solution is to enclose the conveyor or at least provide a windbreak on that side. Should rain, ice, or snow be blown in on one side of the conveyor only, then the result will be a difference in friction on the idlers of that side. This difference may be enough to push lightly-loaded belts off the proper path. Even the difference created when the sun warms one side of a belt in the morning and the other side in the afternoon is enough to cause a belt to wander. Here again, the solution would be some form of conveyor cover.

In some cases, the conveyor's design was not sufficiently strong to withstand lateral winds, and the entire conveyor will wander back and forth in high winds. This is especially true on silo or bunker loading conveyors that have straight vertical supports, as opposed to supports built at an angle. In these cases, contact the conveyor designer or a structural engineering firm for recommendations on reinforcing the structure.

Belt Wander Due To Material buildups

Material accumulations on idlers and pulleys (**Figure 15.5**) will lead to mistracking. When wet and/or sticky materials build up, the accumulations simulate components that are out of round or out of alignment. buildup on a rolling component turns that component into a crowned roller. As the belt tries to find this artificial center, it wanders. (**Figure 15.6**) The differing diameter of the accumulations on different components leads to erratic and unpredictable tracking. buildup can also cause an unequal tension on the belt, which is detrimental to the lives of the belt (particularly with steel cord belting) and of the splice.

Special care must be exercised to keep return rolls,

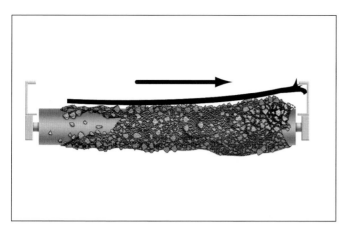

Figure 15.5

Material accumulation on rolling components can lead to belt mistracking.

Figure 15.6

Material remaining on the belt has accumulated on this idler, leading in turn to belt wander.

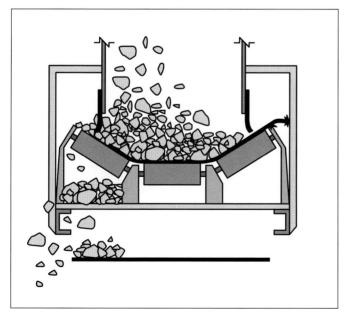

Figure 15.7

Off-center loading can push the belt to the more lightly-loaded side of the conveyor.

snubs, and other pulleys clean. An effective multiple-cleaner belt cleaning system should be installed to prevent material carryback. The transfer points should be engineered, constructed, and maintained to prevent material spillage. If necessary, cleaners can be installed to directly clean snub, take-up, and other pulleys.

Belt Wander from Off-Center or Segregated Loading

Mistracking that arises from loading problems is generally easy to spot, as the belt will run in one position when loaded and another position when unloaded. Of course, this observation may be confused on older conveyors where years of belt training adjustments have altered the natural track of the belt.

If the belt does not receive its load uniformly centered, the load's center of gravity will seek the lowest point of the troughing idlers. This pushes the belt off-center toward the conveyor's more lightly-loaded side. **(Figure 15.7)** This can be corrected by proper loading chute arrangements and the use of adjustable deflectors, grids, and chute bottoms that can be tuned to correct the placement of the load on the belt.

Belt Wander on Reversing Belts

Reversing conveyors, especially older ones or conveyors that run one direction more than the other can be a special source of frustration.

Consider, when a belt is reversed, the tension areas in the belt change location in relation to the drive pulley and loading area(s). Imagine having a conveyor that has a head drive and at the flip of a switch, it becomes a tail drive. When the belt is running toward the drive pulley, the tight side of the belt is on top. When the belt is running away from the drive pulley, the tight side is now on the bottom. This poses some especially difficult problems, as all of the components now contribute differently to the tracking problems. The belt may run fine in one direction and wander all over when reversed, because different sets of rollers and pulleys are controlling belt steering. In order to overcome this type of problem, a survey should be conducted to determine which components are out of alignment and corrections made as required.

Other problems encountered and aggravated by reversing belts relate to off-center loading and different materials on the same belt.

Off center loading can greatly aggravate tracking problems on reversing belts, especially if the load is being applied closer to one end of the conveyor than to the other. This can be corrected by proper loading chute arrangements and the use of adjustable deflectors, grids, and chute bottoms that can be tuned to correct the placement of the load on the belt.

Different materials on the same reversing belt can also cause problems. Suppose the belt has been "tuned" to a specific material with a specific bulk density. Now,

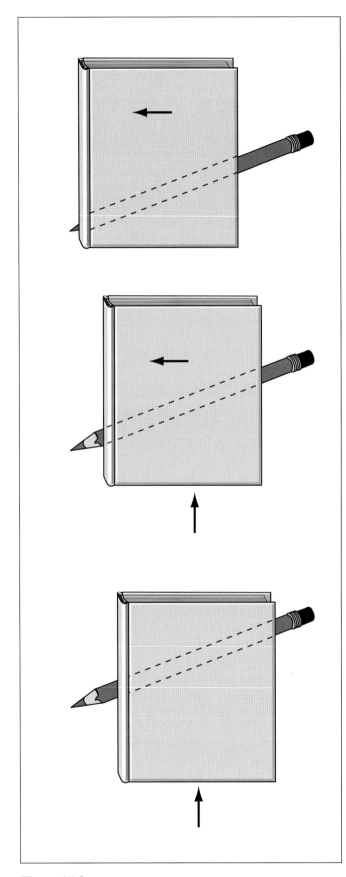

Figure 15.8
The basic rule of belt behavior is that the belt (or book, as demonstrated here) moves toward the side it contacts first.

introduce a material with a different bulk density when the belt is reversed, and all of the training adjustments may go down the drain. Again, in order to overcome this type of problem, a survey of the structure should be conducted to determine which components are out of alignment and corrections made as required.

Basic Belt Behavior

Despite all its various causes, mistracking is still unnecessary. It is a problem that can be controlled or, better yet, corrected. Understanding the basic patterns of belt behavior and undertaking a set of procedures to carefully align the conveyor structure and components and to correct fluctuations in the belt's path can, in most cases, prevent belt wander. This minimizes the expensive problems that accompany mistracking.

Belt behavior is based on simple principles. These principles serve as the guidelines for belt training, the process of adjusting idlers, pulleys, and load conditions to correct any tendency for the belt to run off center.

The fundamental rule of conveyor tracking is simply this: the belt moves toward the end of a roller that it contacts first. If an idler set is installed at an angle across the stringers, the belt will move toward the side it touches first. If one end of the idler is higher than the other, the belt will climb to the high side, because as it lays over the top of the idler, it contacts the higher end first.

This can be demonstrated very simply by laying a round pencil on a flat surface, like a table. If a book is laid across the pencil and gently pushed in a direction away from the experimenter, the book will shift to the left or right depending upon which end of the pencil (idler) the book (belt) contacts first. **(Figure 15.8)**

This basic rule is true for both flat idlers and troughed idler sets. In addition, troughed idlers exert a powerful tracking force. With their troughed configuration, a portion of each belt edge is held aloft at a more-or-less vertical angle. A gravitational force is exerted on that raised portion. If the belt is not centered in the set, the force on the one edge will be greater than the other, pulling the belt into the center of the troughed idler set. This gravitational tracking force is so pronounced that bulk conveyors usually depend upon it as their major tracking influence.

Another constant rule of belt tracking is that the tracking of the belt at any given point is more affected by the idlers and other components upstream (the points it has already passed) than the components downstream (which it has not yet reached). This means where mistracking is visible, the cause is at a point the belt has already passed. Corrective measures should be applied some distance before the point where the belt shows visible mistracking.

With these basic rules in mind, operators and maintenance personnel can begin to make the adjustments to the conveyor that will bring the belt path into alignment.

Checking the Structure

To control belt tracking, it might be best to make a detailed survey of conveyor conditions. This allows measured corrections to be made, rather than to adopt an unplanned, "let's 'tweak' the idlers a little more today" approach. Unfortunately, the traditional methods for the survey and alignment of conveyor components have generally proven unreliable and ineffective.

A traditional method of checking alignment has been to stretch a piano wire from one end of the conveyor to the other and use this wire as a baseline to take the measurements to evaluate alignment. However, this method has a number of potential problems. For example, this wire is vulnerable to shifts in its line. Changes in ambient temperature, from the warmth of the sun, or even the actual weight of the wire itself, can cause the wire to stretch, changing the line. Another–and more serious–problem is that there is no consistently accurate way to measure a 90° angle from the wire. If the wire moves when touched, the accuracy of subsequent measurements is destroyed.

Now, high technology–in the form of laser-generated beams of light set in parallel to the conveyor structure–provides an unobstructed reference for the alignment of conveyor components. (**Figure 15.9**) Laser surveying technology has shown the accuracy required to align all components to protect today's high-speed, high-tonnage conveyor systems.

Laser alignment technology avoids the problems encountered with the old "piano wire" technique. The laser generates an arrow-straight beam out to the limits of the equipment, currently an effective range of 500 feet (150 m), with multiple set-ups allowing unlimited distance.

Use of a laser allows the survey to be taken in daylight or at night. Since a laser beam cannot be touched, it cannot be moved accidentally when taking readings from it. Though sunlight and wind have no noticeable effect on the laser system, high ambient temperatures on long belts may cause the beam to "dance." Greater care must be taken in obtaining measurements under this condition. If the problem is too great, the laser survey may have to be done on cloudy days, at night, or at a time when the ambient temperature is lower.

To check objects set at angles to the baseline, a set of prisms can be used to bend the beam. With a laser transit, the survey crew is no longer trying to measure a perpendicular line; they have created one.

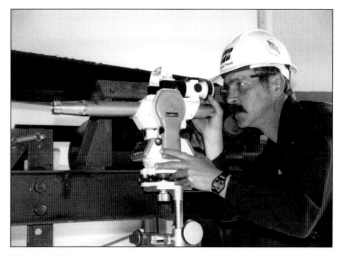

Figure 15.9

A laser transit is used to develop an accurate assessment of a conveyor structure.

Figure 15.10

Specialized equipment is required for the performance of a laser survey.

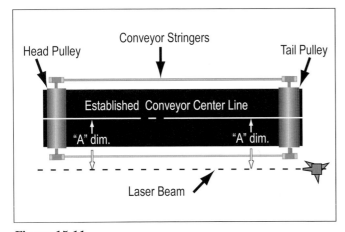

Figure 15.11

Laser Surveying Step 1: Establish a baseline to check that stringers are straight.

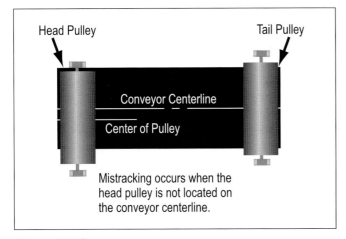

Figure 15.12

Laser Surveying Step 2: Check that the head and tail pulleys are centered on the structure.

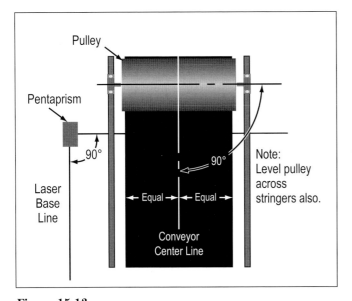

Figure 15.13

Laser Surveying Step 3: Verify that head and tail pulley are perpendicular to the conveyor centerline.

A Laser Survey

To check the alignment of the conveyor and components, a low-power (less than 5 milliwatts) laser transit–resembling a conventional optical surveyor's transit with the addition of a laser tube–is used to "shoot" a baseline along a conveyor. **(Figure 15.10)** Next, a prism is used to bend the laser beam 90° to check the squareness of the installation of the head pulley, tail pulley, drive pulley, and snub pulleys. The elevation of these rolling components is also checked to determine if each is level or if one side is higher than the other. Any rolling component not precisely aligned–in all three planes–can drastically alter the path the belt is following.

The following is the procedure for a laser conveyor survey:

1. A baseline is established to make sure the stringers are in line and the conveyor is centered on the stringers. **(Figure 15.11)**

   The laser's beam of light is first placed to determine the true centerline of the belt conveyor. By measuring from this reference point, the centerlines of the head and tail pulleys can be determined.

   The laser is placed so its beam of light is parallel to the conveyor centerline. This is done by establishing a convenient "A" dimension–the distance from the beam of light to the centerline of the pulleys. This "A" dimension should be the same for both pulleys, within a tolerance of one-sixteenth of an inch (1.5 mm) to ensure the laser baseline's accuracy. All other components will be checked from this baseline.

   From the baseline, measurements are then taken at regular intervals to determine if the conveyor stringers are out of true and, if so, by how far. These intervals are typically approximately ten feet (3 meters).

   If the conveyor stringers were not installed properly or have been damaged, components such as idlers cannot be level, perpendicular, and centered on the conveyor centerline. Measuring from the laser baseline will determine the correct alignment of conveyor stringers.

2. The head pulley and tail pulley are checked to make sure they are on the centerline of the conveyor structure. **(Figure 15.12)**

   The head pulley and tail pulley–the two components that provide the rolling surfaces the conveyor belt is stretched between–must be in perfect alignment to the centerline of the conveyor structure.

   Of course, this does not apply to conveyors with a horizontal curve. In these cases, a precise line has been computed, and the idlers tilted to the radius of the horizontal curve. Any surveying must be done in accordance with the computed design plan for the conveyor's curve.

If the pulleys are perpendicular to the centerline but off center, they will still cause mistracking. It is also possible for both pulleys to be misaligned with the structure. It is usually easier to correct this circumstance by aligning the tail pulley to the head pulley.

3. *The head pulley and tail pulley are checked to ensure they are perpendicular to the conveyor centerline.* **(Figure 15.13)**

From the laser baseline, a pentaprism–a five-sided, light-refracting device that bends the laser beam at a perfect 90° angle–is used to check the pulleys to ensure they are perpendicular to the centerline of the belt within one-sixteenth of an inch (2 mm). **(Figure 15.14)**

4. *Snub, bend, and take-up pulleys are checked to be sure they are centered on the conveyor and aligned perpendicular to the conveyor centerline.* **(Figure 15.15)**

The laser baseline is used to check the remainder of the pulleys in the conveying system. If the pulleys are not placed on the centerline of the conveyor, the belt will run off to one side, possibly interfering with conveyor support structures and risking damage to the belt, structure, or both.

5. *All pulleys must be leveled*

A pulley that is not level will certainly cause mistracking. This may be especially apparent in the conveyor's take-up system.

Many times excess clearance or uneven counterweight distribution in a gravity take-up system will allow a pulley to become tipped, leading to mistracking. These excessive clearances must be removed, and the pulleys leveled and squared to the belt's centerline. This is usually a precision adjustment, as the take-up still must remain free to perform its assigned task of maintaining belt tension.

6. *Carrying side and return side idlers must be installed on the conveyor centerline, must be level, and must be perpendicular to the belt centerline.* **(Figure 15.16)**

Typically, idlers are aligned 90° to the belt centerline. Just as important, idlers should be aligned transversely with the conveyor centerline–they should be centered under the belt. If the troughing idlers are not aligned

Figure 15.14
A pentaprism turns the laser beam at a 90° angle to check pulley alignment.

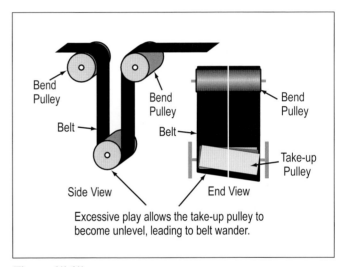

Figure 15.15
Laser Surveying Step 4: Check snub bend and take-up pulleys.

Figure 15.16
Laser Surveying Step 5: Verify that carrying and return idlers are centered, level, and perpendicular to the belt centerline.

uniformly, highs and lows are created in the idler line. These highs and lows will work to steer the belt randomly along the conveyor structure.

The return side of the belt is much more vulnerable to minor idler misalignment. This is because the return side is the low tension side of the belt. The transverse alignment of the flat return rolls is not as critical as that of the troughing idlers, but V-shaped return rolls should be aligned both transversely and perpendicularly with the belt centerline.

Laser Surveys for Traveling Conveyors

"Traveling" conveyors (whether bucketwheel-reclaimer, traveling stacker conveyor, or a tripper conveyor) are greatly influenced by the rail structure on which they ride. For instance, if one rail is higher or lower at a given point than the parallel point on the other rail, the mobile structure can tip or rock (sometimes several inches on tall structures) and cause all sorts of mistracking problems.

Figure 15.17
Benchmarks are applied to the conveyor structure for future analysis of conveyor alignment.

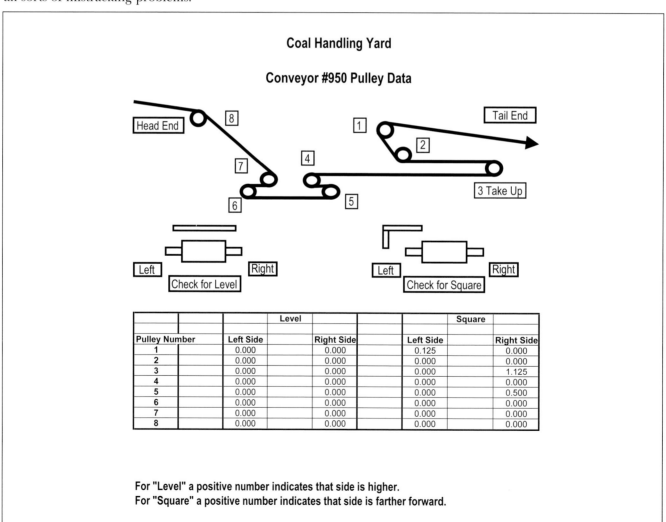

Pulley Number	Level		Square	
	Left Side	Right Side	Left Side	Right Side
1	0.000	0.000	0.125	0.000
2	0.000	0.000	0.000	0.000
3	0.000	0.000	0.000	1.125
4	0.000	0.000	0.000	0.000
5	0.000	0.000	0.000	0.500
6	0.000	0.000	0.000	0.000
7	0.000	0.000	0.000	0.000
8	0.000	0.000	0.000	0.000

For "Level" a positive number indicates that side is higher.
For "Square" a positive number indicates that side is farther forward.

Figure 15.18
Data from a laser survey provides a detailed analysis of conveyor component alignment.

Many times this problem is overlooked when trying to find the cause of belt mistracking and subsequent damage. The "traveler" part of the system might be parked in an area where the rails are level when a survey is performed. The survey results then would show everything to be in alignment, but when the traveling system is moved to a different location, the belt mistracks due to the supporting structure not being level.

If the structure shows signs of belt damage or structural damage in line with the belt edges, the rail system should also be laser surveyed to determine the high and low points in the system. In most cases, the rail system does not have to be exactly level along its length, but the two rails must be level side to side at each specific point.

Rail systems built on short ties on unstable ground or carrying very heavy structures are especially vulnerable to this problem. Heavy train travel close to the support rails may also have an undesirable effect as the vibration from the trains may be transmitted through the ground and cause localized settling of the support rails. In order to overcome this type of problem, a laser survey should be conducted to determine which components are out of alignment and corrections made as required.

The rail systems must also be checked for parallel alignment. Improper alignment may cause the carrying wheels to "ride up" the inside or outside of the rail, causing the same effect as one rail point being higher than its opposite counterpart. In severe cases, the carrying wheel may ride up to the point of derailment, leading to severe structural and belt damage.

Analyzing the Survey

The next step is to inscribe a permanent series of benchmarks or alignment points at regular intervals along the conveyor structure. These marks are used as standardized measurement points when adjusting the existing components or installing new ones. **(Figure 15.17)**

All of this information is recorded and computer analyzed to provide both a digital and graphic print-out of the conveyor alignment. **(Figure 15.18)** The information tells which components are out of alignment, and by how much, so the plant maintenance crew or a contractor can adjust these components to improve the belt's tracking. At the same time, recommendations are also made as to how to improve the conveyor system through the installation of cleaning, sealing, and belt training equipment.

Upon receipt of the report, the plant personnel can make accurate adjustments–shifting rolling components fractions of an inch in the right direction to correct misalignment or to compensate for twisted stringers and other problems.

Once the survey and the follow-up alignment procedures are completed, the work is done. By doing repeat surveys of the same conveyor at regular intervals–annually, for example–plant management can determine if the condition of the conveyor structure is becoming progressively worse or if other conditions–such as subsidence of the ground under the conveyor–are occurring. This information can be used to prevent unexpected shutdowns and subsequent loss of production by alerting the plant engineering and maintenance staff to problems as they develop.

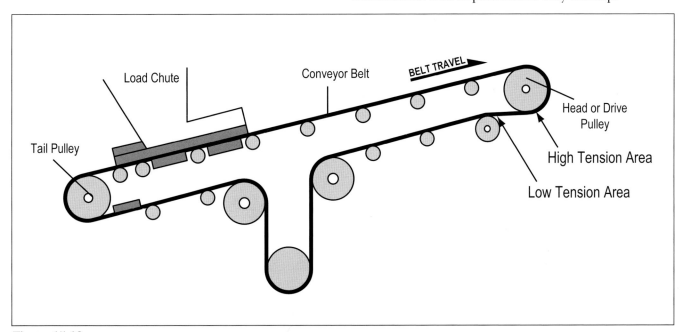

Figure 15.19

Belt tension is highest at the conveyor drive and lowest immediately after the drive pulley.

Laser surveying for conveyor alignment is a good example of adapting an advanced technology into a new application. It provides a precise method to learn which corrections to make to improve conveyor belt tracking. For that reason, it should become basic to every conveyor installation and maintenance situation.

Training a Conveyor, Tracking the Belt

The terms training and tracking have often been used interchangeably. But in the interest of precision, we are going to elevate the language and give each term a specific meaning.

Training the conveyor is making permanent adjustments to the structure, the loading equipment, and the rolling components to get the conveyor to run straight. Tracking is a dynamic process that will readjust the belt while it is runing. Tracking uses various devices that will note when and how far the belt is off and then, self-aligning to steer the belt back to the correct path.

Once the structure and pulleys are properly aligned, you can turn your attention to getting the belt to run true. Getting the belt to run in the center is a process of adjusting idlers and loading conditions to correct any tendency of the belt to run outside the desired path. This is usually referred to as training the belt.

The belt is most often trained by adjusting the idler(s) near the point where the belt runs off-track. The path of the belt across a given idler set is affected more by the idler previously crossed than by the idler that it will cross next. The largest tracking deviation does not occur at a misaligned idler but farther downstream. Consequently, the adjustments to idler alignment should be spread over some length of the conveyor preceding the area where trouble is visible. The movement of one idler generally has its greatest training effect in an area within 15 to 25 feet (5 to 8 meters) downstream of that idler.

A Procedure for Training

The following is a step-by-step process for training the belt.

1. *Determine Locations (and Causes) of Mistracking*
    Of course, the first step to improving the tracking of an existing conveyor is to determine the locations of mistracking and the causes of that wander. Remember that mistracking is caused by problems or components situated prior to the point where the mistracking is visible.
2. *Correct the Problems Causing Wander*
    Just as the first step is identifying the locations and then the causes of wander, the second step is remedying those causes of mistracking. That way, the belt wander may be cured without extensive "tweaking" of the conveyor.
3. *Determine the Areas of Belt Tension*
    The key to knowing where to start the training procedure is to determine the regions of high and low belt tension. That is because adjustments in low-tension areas of the belt have the highest impact on the path of the belt. In high tension areas, there is too much pressure on the belt for the relatively minor idler movements to have significant impact on the belt path.
    Belt tension is highest at the drive pulley and lowest immediately after the drive pulley.
    (Figure 15.19) By identifying and starting in the low tension areas, the training process can have the greatest impact with the least amount of changes.
4. *Track the Belt*

> *Before making any corrections, lock out the conveyor power.*

Starting from the low tension areas, move around the conveyor, making adjustment to center the belt path. Start immediately after the drive pulley, whether that is at the discharge, center, or tail of the conveyor.

Adjust the first idler set before the place where the belt is visibly off-track by turning the idler in the direction opposite the misalignment. Then, restart the conveyor and check for belt misalignment. Run the belt to evaluate its effect, but wait for at least two or three complete belt revolutions before making further adjustments, as the effects of idler movement are not always immediately apparent. If the belt is still off-track, lock it out and adjust the idler immediately before the previously adjusted idler set, making slight adjustments rather than major ones. It is best to shift only one idler at a time, as shifting additional idlers may cause over-correction. If the check shows the belt's path has been over-corrected, it should be restored by moving back the same idler rather than shifting additional idlers.

First track the belt empty, all the way around the conveyor, making especially sure that the belt is centered into the loading zone first, and the discharge area secondly. Note that as you correct in one area, you may be creating additional problems to "undo" in a later area.

Continue this procedure until the belt tracks up the center of the idlers at this point on the conveyor. Repeat this procedure at other points along the belt, if necessary, until the return side runs true. Then follow with the carrying side of the belt, moving in the direction of belt travel from load zone to discharge point.

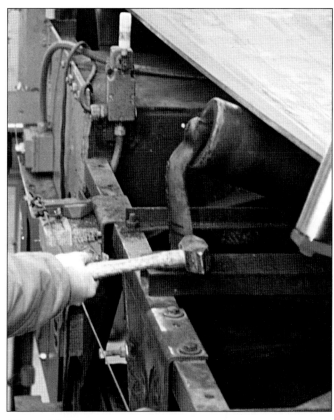

Figure 15.20
"Knocking the idlers" pivots the base of these rolling component to steer the belt path.

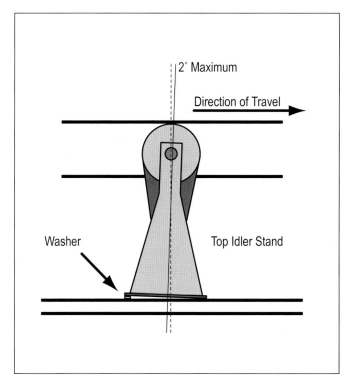

Figure 15.21
Tilting the idlers with a shim such as a washer is one technique to improve belt training.

Start with the belt empty; after training is completed, run the belt loaded to check the results. If the belt runs true when empty and mistracks when loaded, the problem probably comes with off-center loading or from accumulations of material creating off-center rolling components that now have added training effect because of the increased weight on them. Start with a light load and work up to the load that the conveyor was designed to handle. Before the belt is put on a full production schedule, all training adjustments should be finalized.

5. *Record the Corrections*

It is critical that records be kept of noting the problem areas and detailing the corrections made. That way, you will not correct, re-correct, over-correct, and counter-correct the solutions. When problems return repeatedly to an area, it will indicate that the original cause was not identified and corrected.

"Knocking" the Idlers

The most basic training technique is to adjust idlers. Training a belt by using its troughing idlers is accomplished by shifting the idler axis with respect to the path of the belt. This is commonly called "knocking the idlers." **(Figure 15.20)** Training a belt by shifting the position of one or more idlers is the same as steering a bike with its handlebars. When you pull one end of the handlebars (or the idler) toward you, the bike (or belt) turns in that direction. This is in keeping with the basic rule of belt training: the belt will steer to the side of the idler it touches first.

This handlebar principle of steering is a sound one, but only if the belt makes good contact with all three troughing rollers. So before training a belt, it should be checked to be sure it is troughing well at all points along the carrying strand, even when unloaded. If the belt does not "sit down" in the trough, there may be a problem with its compatibility with the structure. In this case, the belt is likely too thick and therefore not suitable for the given trough, and so that belt may never track correctly.

If the belt is mistracking toward one side, moving that side of an idler forward will provide a correcting influence on the belt's path. It is best to do this one or two idlers in advance of the point of mistracking. Again, it is better to adjust several idlers a small amount than to move one idler a large amount.

Obviously, such idler shifting is effective only for one direction of belt travel. A shifted idler that has a correcting influence when the belt runs in one direction will likely misdirect the belt when it is running in the other. Hence, reversing belts should have all idlers squared to the structure and left that way.

Shifting the position of idlers has benefits in belt training. However, there are drawbacks as well. It will be recognized that a belt might be made to run straight with half the idlers knocked one way and the other half knocked in the opposite direction, but this would be at the expense of increasing rolling friction between belt and idlers. Idlers turned in all directions in an effort to train the belt create extra friction, resulting in increased belt cover wear and increased power consumption.

Adjustments should be made to idlers only, and never to pulleys. Pulleys should be kept level with their axis 90° to the intended path of the belt.

Tilting the Idlers

It may be desirable to place extra pressure on carrying idlers where a strong training influence is required. One way to accomplish this is to raise these idlers slightly above the line of adjacent idlers. This can be accomplished by tilting the idler slightly forward in the direction of belt travel. **(Figure 15.21)**

However, too much tilt can reduce the area of contact between belt and roller or cause excessive wear on the inside of the belt. It is best, therefore, not to move the uppermost point of the roller more than two degrees—generally 1/8 to 3/16 of an inch (3 to 5 mm) higher. The simplest way to accomplish this is by inserting a shim, which can be as unsophisticated as a steel washer, underneath the rear leg of the idler frame.

This method has an advantage over "knocking" idlers, because it will correct erratic belts that track to either side. It has the disadvantage of encouraging belt cover wear through increased friction on the troughing rolls. Consequently, this technique should be used sparingly, especially on idlers with the higher angle of trough. The effect of tilting idlers is counterproductive on reversing belts. Return idlers should not be tilted.

Some suppliers manufacture idlers with a built-in forward tilt; care must be taken when installing these idlers to make sure they are leaning in the direction of belt travel.

Raising Alternating Idler Ends

Another method to center the belt as it approaches the tail pulley is to slightly advance and raise alternate ends of the return rolls nearest the tail pulley. **(Figure 15.22)** The theory here is that the deliberately induced mistracking in opposite directions produces competing forces that work to center the belt. Though this may sound reasonable in the abstract or the laboratory, the practical application of it is problematic. Why inject instability into a system where the goal for optimum operations is stability? It could be argued that there is enough of a problem with getting the system square in order to run true, without adding two more variables in the form of deliberately-misaligned idlers.

Training at New Conveyor Start-Up

If the conveyor system has been designed and built in accordance with good engineering and installation practice, the belt will probably track at start-up on a path close to the desired one. There may be minor variations from the ideal structure that result in the belt not tracking perfectly; however, in these circumstances the variations should be relatively minor, so the belt can be operated without damage long enough for a training procedure to take place.

The first movement of the belt through a new conveyor should be slow and intermittent, so any tendency of the belt to mistrack may be quickly recognized, and the belt stopped before damage occurs. The first corrections must be those at points where the belt is in immediate danger of being damaged. Once the belt is clear of danger points, the conventional sequence of training operations, as noted above, can be followed.

Insufficient attention at start-up can create problems, including serious runoff and edge damage, belt fold-overs and creasing, spillage, and damage to other conveyor components. For conveyor start-up, observers should be positioned at locations where trouble might be expected, or the belt is at greatest risk. These "spotters" should have a radio or telephone, or at minimum, have a pull-rope emergency stop switch within easy reach.

In severe cases it may be necessary to shut the conveyor down, make any adjustments indicated, and then restring and reposition the belt before a new start-up is undertaken.

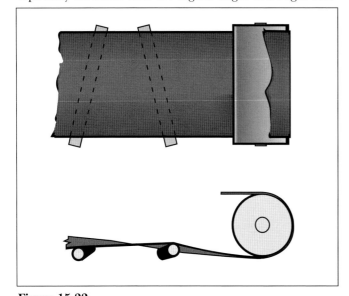

Figure 15.22

Raising the opposite ends of return rollers immediately before the tail pulley can improve belt alignment.

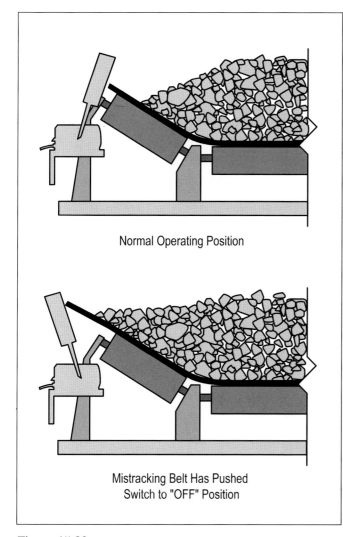

Figure 15.23
Belt misalignment switches will shut off conveyor power when activated by a mistracking belt.

Figure 15.24
Rollers are installed on either side of the belt to serve as vertical edge guides.

Training of Replacement Belts

A new belt–whether new belting on a new conveyor or a replacement belt on an established system–often has to be gradually "worn in" like a new pair of shoes. It is relatively rare to pull a new belt onto the conveyor, splice it together, push the conveyor start button, and have it track down the middle of the structure.

Remember, the conveyor may not be neutral to the new belt, particularly in the case of a new belt going onto an existing conveyor. If numerous training adjustments were made over time to accommodate the mistracking of the previous belt, these adjustments may have to be "undone" to allow the new belt to track correctly.

Some new belts will tend to run to one side in one or more portions of their length, because of a temporary unequal distribution of tension arising from the storage, handling, or stringing of the belt. In most cases, operation of the belt under tension will correct this condition. Loading the belt to 60 percent capacity will help the belt fit the conveyor in the same way as running a new car for a "break in" period acclimates the system to the application.

Mechanical Training Aids

There are occasions when the tracking procedure is not successful at providing a long-term solution to a belt wander problem. So the plant is faced with repeating the tracking procedure on a frequent (sometimes daily) basis. Or they install some active tracking system that will automatically induce corrections and steer the belt into the correct path while the belt is in operation. These systems range from simple to complex, from passive protective equipment to active belt steering systems.

Belt Misalignment Switches

One mechanical tracking aid that is a requirement for efficient conveyor operation is a system of belt misalignment switches. **(Figure 15.23)** These are installed at intervals along the length of the conveyor near the outer limit of safe belt travel on both sides of the belt. When the belt moves too far in either direction, it contacts the switch's roller or lever, pushing it past its limit and interrupting the conveyor's power circuit. The belt stops and the operator has the opportunity to realign the belt. In many cases he will need to walk the conveyor to manually reset the switch before operation can begin again. Some devices have the desirable ability to send multiple signals–one indicating an alarm representing a pre-set amount of belt wander, and the second signal cutting off power due to a more serious tracking problem.

Of course, the tripping of a belt misalignment switch is a signal something is wrong with the conveyor system.

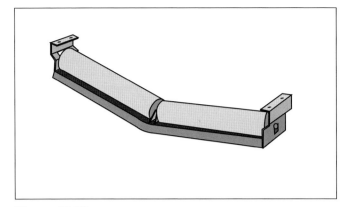

Figure 15.25

V-Return idlers can assist in keeping the belt in the center.

It is like a light on a car dashboard that shows red when the engine is too hot. It is possible to ignore this light–to merely reset the switch and resume conveyor operations–but both the car's warning light and the conveyor misalignment switch should serve as an alarm that there may be a more serious, more expensive, possibly catastrophic problem. Conveyor stoppages can be a nuisance and are, in fact, very costly; each outage creates downtime and lost production. The answer is to find systems that will do more than shut off the conveyor; systems that work to prevent or cure misalignments.

Vertical Edge Guides

One such system places spools or rollers, supported by simple frames, a few inches from the belt edge. **(Figure 15.24)** These vertical edge guides are installed in a position approximately perpendicular to the belt's path to keep the belt edge away from the conveyor structure. These side guides do not train the belt. Rather than preventing belt wander, they perform a damage control function, allowing the belt to strike a rolling surface rather than the fixed, unyielding structural steel. Vertical edge guides should be installed so they do not touch the belt edge when it is running normally. If they contact the belt continually, even though they are free to roll, they can wear off the belt edge and cause ply separation. Vertical edge guides can lead to severe belt or structure damage when used to compensate for persistent misalignment problems. It would be a far better solution to locate the source(s) of the training problems and correct them.

V-Return Idlers

Another hardware addition that can help remedy belt wander is to install V-Return idlers on the return strand. V-return idlers are becoming more popular on longer, high-tension conveyors. They are available in two versions, traditional V-Rollers **(Figure 15.25)** and

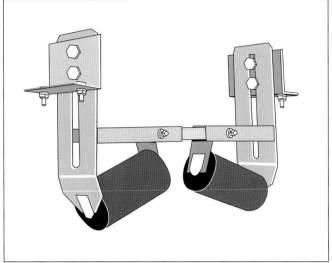

Figure 15.26

Inverted V-Return Idlers can drive the belt into alignment.

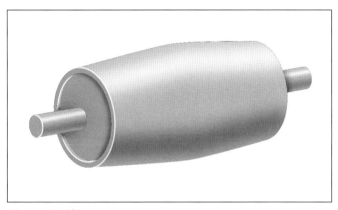

Figure 15.27

Crowned pulleys are slightly larger in the center than on the ends to steer the belt.

inverted V-Rollers **(Figure 15.26)** Both systems form the belt into a trough to assist in steering it into the center. They rely on brute force to muscle the belt back into line and so tend to place added stress on the belt, which can lead to damage. These systems are more expensive and require somewhat more maintenance than a conventional return idler. Care must be taken to keep material from building up on the linkage between the two rolls.

Crowned Pulleys

Pulleys that have larger diameters at the center than at the edges are sometimes used to provide a centralizing effect. These "crowned" pulleys **(Figure 15.27)** operate from the basic principle of tracking that the belt moves to the part of the component it touches first. In the case of crowned pulleys, the belt first touches the higher middle area of the pulley and so is directed into the center. Crowned pulleys are most effective on conveyors with short, low-

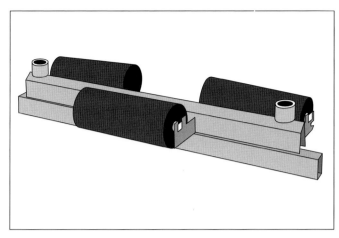

Figure 15.28

The in-line sensing roller belt tracker pivots on its central bearing to keep the belt on the correct path.

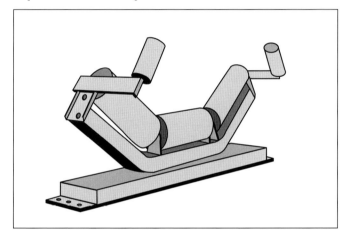

Figure 15.29

On the leading sensing roll tracker, the belt passes the sensing rolls before it reaches the pivoting idler.

tension belts. With higher tension, steel cable belts, and troughed conveyors, little steering effect is obtained from the crown of the pulley. Crowns are most effective where there is a long unsupported span–four times belt width or greater–approaching the pulley. As this is not often possible on the carrying side of the conveyor, the use of crowned head pulleys is relatively ineffective and may not be worth the stress it produces in the belt. They are somewhat more effective when used as a conveyor's tail pulley.

Dynamic Tracking Solutions

If everything about the conveyor is aligned and leveled, and if the belt is troughed and trained, no self-aligning idlers are needed. However, rarely are imperfections entirely absent. Most conveyors need some tracking correction. Belt tracking devices are especially useful near the head and tail pulleys, since these are the loading and discharge areas where the belt should always run on center to prevent spillage and damage. Providing one or two training idlers at intervals in advance of each pulley will guide the belt into these pulleys.

These belt tracking systems are designed to "self-align." That means the force of the mistracking belt causes the idler to repositionitself to create a force or a steering action that directs the belt back into the center.

Many self-aligning idlers carry the seeds of their own destruction. Because they are designed to pivot and so provide a correcting influence on the belt path, they are particularly vulnerable to the accumulations of fugitive material. Piles of spillage can block their range of motion or seize the pivot bearing. This can lock the belt tracking idler into a position where it functions as a "misalignment" idler. It now pushes the belt out of the proper track, creating (or worsening) a problem it was installed to correct. To correct the now misaligned system, the maintenance crew ties the belt tracking idler into (approximately) the right position. In such cases–if a self-aligning idler were not capable of functioning properly–it would be better to remove it, rather than to just "tie it off."

These tracking aids–like self-aligning idlers, edge guides, and wander switches–are never totally effective in controlling belt wander. All these systems work under the disadvantage of being "after the fact"–they correct mistracking after it has occurred. A certain amount of wander must happen before the required correction can take place. But these systems act as a form of insurance against a problem becoming so severe the belt suffers costly damage before the mistracking is discovered.

Inline Sensing Roll Trackers

The simplest tracker design has a carrying roll held in a framework that is mounted on a central pivot bearing. The guide rolls are mounted on either side with their centerline running through the idler's pivot point. **(Figure 15.28)** Movement of the belt against either sensing roll creates friction and causes that sensor to move in the direction of belt travel. This pivots the entire idler. In keeping with the basic rule of tracking that the belt always moves toward the side it contacts first, the pivoted idler steers the out-of-track belt back to the proper path.

Inline sensing rollers have almost no leverage. They require considerable friction force from the belt edge to cause correction. With this idler design, the belt wanders from side to side; the correcting action is caused by the belt literally slamming into one side or the other. When the correcting action takes place, the idler may "kick over" with such force that the belt is then directed all the way over to the other side of the structure, in turn contacting the roller on the other side of the tracking idler and correcting the belt path in the other direction. The belt can be kept constantly in

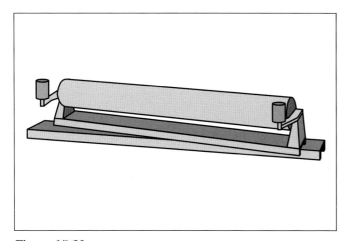

Figure 15.30

Pressure against the sensing rolls pivots a return tracking idler to correct the belt path.

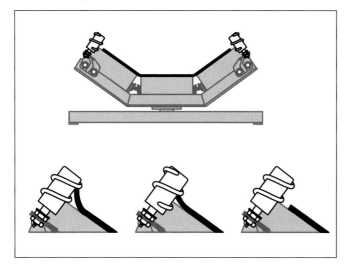

Figure 15.31

Spiral wrapped sensing rolls can improve the performance of some belt tracking idlers.

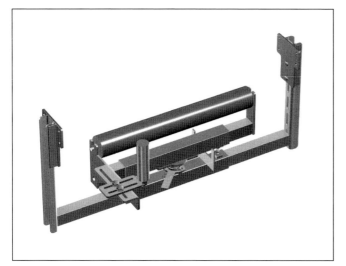

Figure 15.32

Designed for use on the conveyor return, the torsion spring tracker requires only one sensing roll.

motion, back and forth between the two sides, risking edge damage and overuse of the bearing. Because the tracking idler has a single, central pivot point, belt movement to one side brings the opposite guide roll into a hard, pinching contact against the belt, which can lead to edge damage.

Leading Sensing Roll Trackers

The most common tracker design has a carrying roll (or troughing set) held in a framework that is mounted on a central pivot bearing. Guide rolls are mounted on short arms and positioned one to three inches (25 to 75 mm) from the belt on both sides in advance of the pivoting roller. Some are designed for use on the upper or carrying side of the belt (**Figure 15.29**); others are for the lower or return side. (Figure 15.30)

Movement of the belt against either guide roll causes the steering idler to pivot, correcting the belt path back toward the center. In keeping with the basic rule of tracking that the belt always moves toward the side it contacts first, the pivoted roll steers the out-of-track belt back to the proper path. These self-aligning idlers serve as a limit switch, correcting the tracking when the belt has moved out of the desired path.

Sensing rollers installed on short arms in advance of the steering idler have slightly more leverage than the inline sensor idlers. But they still require considerable force from the belt edge to cause correction. This tracker design suffers from all the delay, pinching, and fugitive material problems of the inline sensing idler.

The leading sensor roll tracker is the most popular and most common tracking idler. It is supplied on almost all new conveyors sold, and installed at intervals of approximately 100 feet (30 m) on both the carrying and return sides.

In the field, however, these trackers are commonly seen in two unsatisfactory conditions. The first condition is "frozen" from material accumulations or corrosion of the center pivot. This problem can be solved with better maintenance, or a higher quality pivot point.

The second condition is "tied off" that is, locked in place with a rope or wire, so the tracker provides the equivalent of a "knocked" idler. The reason these are tied off originates in a design flaw.

The sensor rolls swing in an arc about the center pivot and, therefore, must be spaced far enough apart to not pinch the belt when the rolls reach extreme positions. As the pivot becomes bound up from lack of maintenance or corrosion, the idler will not react until the belt has mistracked a distance equal to this spacing. In essence, the idler is designed to over-steer, and, therefore, becomes an unstable mechanical control system. They overreact, providing unpredictable results, and so they are "tied off."

In short, the problems caused by these leading sensing roll trackers are worse than the problems they were supposed to cure. In most cases, the plants would be further ahead by eliminating the leading sensor roll trackers from their conveyor specifications, and spending their money on more precise and stable tracking systems.

Spiral Rolls

Some leading sensing roll trackers achieve improved performance with the use of spiral-wrapped sensing rollers. **(Figure 15.31)** When installed on the center-pivot belt trackers, these rollers spin when touched by a wandering belt. As the roller spins, its spiral wrap pushes the belt down onto the tracking idler. This improves the reaction speed and force of the tracking idler's correction.

Torsion Spring Tracker

The torsion spring tracker is an improved version of the leading sensor roll tracker. **(Figure 15.32)** This system removes one sensing roll and incorporates a spring into the pivot. **(Figure 15.33)** This spring keeps the one sensor in contact with the belt edge at all times. As the belt mistracks in either direction the idler will compensate by pivoting and steering the belt. One drawback of this tracker is the fact that it does not function with a carrying side troughing set.

Spring-loaded leading sensor trackers tend to have the sensing rolls installed on longer arms in advance of the steering idler. This creates more leverage and a greater mechanical advantage in converting belt wander into steering torque. There is no delay in reaction of this tracker due to the fact that the sensing roll is in constant contact with the belt. There is also no pinching as there is only one sensing roll. Because of the constant motion of the idler, fugitive material will have a much harder time in freezing the bearing.

Multi-Pivot Belt Trackers

There is another belt tracking system that uses the force of the wandering belt to position a steering idler and so correct the path. This device uses a multiple-pivot, torque-multiplying system to supply a mechanical advantage to improve tracking correction. **(Figure 15.34)**

Figure 15.33

A spring in its pivot point keeps this torsion spring tracker in contact with the belt edge.

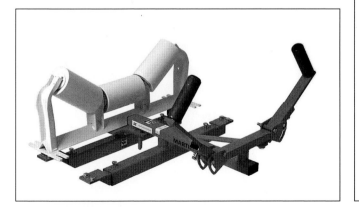

Figure 15.34

This multi-pivot belt tracker is available in units for the conveyor's carrying side (shown here) or return side.

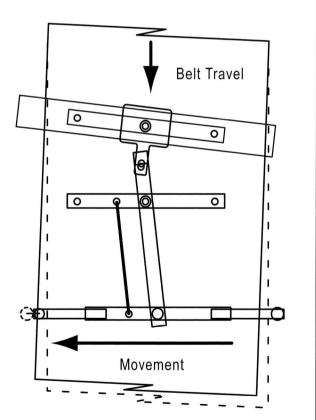

Figure 15.35

The torque multiplying system of the multi-pivot belt tracker provides substantial mechanical advantage to steer the belt.

This style of belt tracker transfers the motion of mistracking to the steering idler through a unique parallel linkage. This requires less force to initiate the work and, as it steers, needs less force to turn the belt. Belt tracking becomes a continuous, active, precise micro-tuning of the belt path.

This tracking device uses guide rolls that are set very close–1/4 inch (6 mm)–from the belt. With the rollers set at the edge of the belt, the device can sense smaller movements of the belt and make corrections after very slight misalignments. Rather than waiting for a powerful mistracking force, the multi-pivot belt tracking device adjusts constantly, reacting to small mistracking forces, providing continuous, precise corrections of the steering roller.

The side guide sensing rollers use longer arms to increase the distance from the guide rolls to the steering idler. This allows the unit's torque arm to act as a force multiplier, increasing the mechanical advantage of the steering action. As a result, the belt tracking system can correct the belt line with one-half the force required for conventional tracking idlers. **(Figure 15.35)**

Unlike the other belt trackers, the multiple-pivot tracking device is installed so the belt crosses the steering roller before it reaches the guide rollers. This means the guide rolls adjust the "corrected" belt path rather than the mistracking belt path. The result is a roller that is continuously working to prevent the belt from moving very far from the proper path **(Figure 15.36)**

The multi-pivot design allows the rollers to move perpendicular to the structure's centerline, while directing the steering idler to the proper angle instead of pivoting and pinching the belt edge.

The continuous movement of the multi-pivot belt-tracking device prevents a buildup of material from seizing it in a misaligned position. Consequently, it provides a consistent and reliable belt tracking system.

Variations of Multi-Pivot Belt Trackers

Several manufacturers have created a slight modification to the multi-pivot belt tracker. These devices use the same force amplification geometry, but the idler itself slides laterally as well as pivoting. In a sliding idler situation, the sensing roll has to overcome the resistance to pivoting as well as the friction force of trying to move an idler from under a belt. This greatly decreases the overall steering force of this tracking system.

Trackers For Reversing Conveyors

Conveyors that run in two directions have always been the "no-man's land" of belt tracking. With these reversing conveyors, even experienced plant personnel are hesitant to adjust the idlers and perform the other maintenance "tricks" typically used to train wandering belts. And conventional belt training devices cannot be used, all for the same reason–because what works to centralize a belt's path when it runs in one direction often has the opposite effect when the belt direction is reversed. A pivoted idler that correctly steers the belt when it is operating in one direction, will work to mistrack a belt moving in the opposite direction.

Recently, a few manufacturers have developed trackers for reversing belts. The inline sensing roll

Figure 15.36
With the multi-pivot tracker, the belt crosses the steering roller before it reaches the guide rollers.

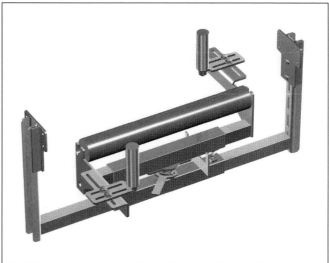

Figure 15.37
This torsion spring tracker has been fitted with a second sensing roll for use on reversing belts.

trackers will correctly steer these belts as the sensors are not direction dependant. The torsion spring tracker can be modified to accommodate reversing belts. **(Figure 15.37)** By adding a second arm and sensor in the opposite direction, the tracker will switch sensing arms based on the direction of the belt movement. Any of these reversing trackers will have all of the benefits and shortcomings associated with their use on one-direction conveyors.

Installation of Belt Trackers

When installing a self-aligning idler, the center roll is typically elevated one-half to three-quarters of an inch (12 to 18 mm) higher than the rolls of the adjacent conventional idlers. This increases the belt pressure causing the corrective action of the tracker to be more effective. This is applicable to both troughed (carrying side) self-aligners and flat (return side) idlers. Some training idler manufacturers build this feature into their various models.

Another technique to improve the performance of belt tracking systems installed on the belt's return side is to reinstall the conventional return idlers on either side of the tracking idler so the belt runs under them rather then over them. This increases the pressure of the belt on the tracking idler, allowing it to work more effectively.

Rubber-covered rollers are often useful on belt tracking devices, particularly where the material is slippery or the belt wet from the climate or the process. These rollers will require replacement more often than "steel can" rollers, but are often necessary to achieve the friction needed for belt tracking.

It has been calculated that the steering force of tracking idlers is most effective at small angles. If the angle is too great–over roughly 5° of misalignment–the belt slides across the idler like a car hydroplaning across a patch of ice. Consequently, it is more effective to stimulate quick, low-angle corrections of belt mistracking than to wait for one larger angle. All trackers will eventually reach this limit.

Power Consumption and Belt Trackers

Any alteration in the conveyor equipment, from "knocking" idlers to installing special mechanical tracking devices, has implications for the conveyor's power requirements. CEMA notes that if idlers with a degree of tilt greater than two degrees are used throughout the conveyor, the additional power requirements can be considerable and should be calculated. CEMA also recommends that if "V" return rollers are used, the idler resistance rating should be increased by five percent.

Similarly, the power requirements of belt tracking idlers should also be considered in the engineering of belt conveyors. As noted above, there are several styles of tracking idlers, all designed to exert a centering force on the belt perpendicular to its travel. The component of this centering force in the direction of belt travel must be considered in the power consumption of the conveyor.

Analyzing the power consumption of a tracking idler requires knowledge of the load on the idler. This load is due to the weight of the belt and any component of the belt tension arising from the idler misalignment. It is common practice to install tracking idlers one-half to three-quarters of an inch (12 to 19 mm) above the standard idlers. This results in greater load on these idlers, which creates enough centering force to influence the travel of the loaded belt. This extra load is described by CEMA as Idler Misalignment Load (IML).

In operation, the typical tracking idler can pivot from two to five degrees. For an idler that pivots 3.5 degrees, it has been calculated on an 18-inch (450 mm) belt with 100 PIW slack side tension and a coefficient of friction of 1.0, that the centering force is approximately 110 lb. (50 kg). [Tr = PIW x BW x tan 3.5]. The component of the centering force in the direction of belt travel would be approximately 6.5 lb. (3 kg). [T = sin 3.5 x Tr]. On the other end of the belt size spectrum, a 72-inch (1800 mm) belt with 500 PIW slack side tension would have a centering force of approximately 2200 lb. (1000 kg) and a component in the direction of travel of approximately 60 lb. (27 kg).

Using the formula HP = T x FPM/33000, it can be determined that if the 18-inch belt was traveling 200 feet per minute, it would require .05 hp (.04 kW) for operation of a tracking idler with a coefficient of friction of 1.0. Traveling at 700 feet per minute, the 72-inch-wide conveyor would require 1.25 hp (0.9 kW) to compensate for the drag from a tracking idler. This additional power requirement should be multiplied by the number of tracking idlers installed.

If the tracking idler becomes frozen and is neither rotating nor pivoting, it can add a substantial power requirement. This could be estimated by taking the weight of the belt and the components of the slack side tension on the idler and estimating a friction factor. Multiplying the load by the friction factor will give an estimate of the tension that can be used with the belt speed in the formula [HP = T x FPM/33000] to estimate the power consumption of the frozen tracking idler.

Chapter 16

Considerations for Specific Industries

While there are many things that hold true for all belt conveyors handling bulk materials regardless of industry, there are numerous factors that should be considered for each specific industry. These relate to the materials, conditions, equipment, and standards found in the industry.

The following are capsule looks at some of these industry-related issues. While these observations will hold true in most situations, they may or may not apply to any specific operation.

Industries:
- Aggregate, Crushed Stone, Sand & Gravel
- Bulk Transportation
- Cement
- Coal Mining (Underground)
- Foundries/Metalcasting
- Metal Mining & Production
- Power Generation (Coal-Fired)
- Pulp & Paper
- Surface Mining

> "... each industry has a number of specific factors that should be evaluated."

Aggregate, Crushed Stone, Sand & Gravel
In General
- With the price for aggregate products at a low price per ton, cost containment is a common theme in this industry. Conveyor improvements will need to reflect this low overhead by providing

obvious cost-effectiveness and a prompt payback.
- In many areas, permits are hard to obtain. This means existing pits must be fully exploited while commercial and residential development fills in around them. This makes concerns like dust, spillage, and noise more important.

Material
- One advantage in this industry is that the raw material handled tends to be more consistent due to the minimal processing required for the end product. This allows for more plant-wide standardization of accessories than in most industries.

Conveying
- Plants vary greatly in size and sophistication. The belts tend to be in poorer condition than most industries, and they may contain used belting and lots of mechanical splices. The size of equipment on the production side of the plant may depend on the in-feed size of the crusher. In the quarry, conveyors are larger and in the medium to heavy-duty range.
- Impact can be severe under truck dumps and crushers; it can be light on finished product belts. The use of impact cradles is common, but careful attention to proper duty ratings is essential.
- Chute plugging is a common problem in the screening areas of the plant. Air cannons or vibrators are an effective means to reduce these blockages.
- Spillage of large rocks presents hazards to tail pulleys, and so winged pulleys are commonly used as tail, bend, and in some cases head pulleys. These create belt flap, making cleaning and sealing difficult. The installation of return belt plows and the wrapping of wing pulleys can solve these problems.
- Gravity take-ups are often built up with spilled material. This causes excess tension and uneven operation. Take-up frames are often loose to prevent binding, but this leads to belt mistracking. Take-up frames can be rebuilt to remove the slop, and "dog houses" can be installed over the take-up pulleys to prevent spilled material getting into the pulley slides.
- Skirtboard seals are often rudimentary; the use of used belting as sealing strips is a common practice. Chutework is often worn, rusted, or flimsy, and wear liners are located high off the belt, making the installation of engineered skirtboard seals a challenge. However, because of the generally open construction of these conveyors, rebuilding the skirts is generally easily accomplished. Rebuilding the skirtboard-and-seal system to the standards discussed in this volume can control most sealing problems. Self-adjusting skirting works well.
- Belts frequently mistrack due to poor loading practices and the use of water in the plant. The installation of multi-pivot training devices before the tail pulley and after the load zone can control most problems.

Belt Cleaning
- Fine crushed material on wet belts can be very difficult to clean, as the material can stick to the belt like a suction cup. Often particles are hard and sharp, which causes them to wedge between blades and the belt. Certain types of sand are also very sharp and abrasive.
- Using softer urethanes for blade construction, combined with water sprays to help remove the material, can reduce the wear on the pre-cleaners.
- In some cases, hard-metal secondary blades wear in an uneven "castellated" pattern. This is caused by small particles wedging between the belt and blade and allowing other particles to stream through. In some cases, the wear is accelerated by a slightly acidic quality of the water. This wear can be reduced through the use of tungsten carbide blades and water sprays. Frequent flushing of the cleaner with a water hose is also beneficial.

Dust Management
- Spray-applied water is the typical dust suppression method. The higher rate of application for plain water increases problems with material handling, such as increased carryback or buildup on screens. In addition, some state and local governments have restricted the use (and disposal) of water. As a result, many plants are now using surfactant or foam suppression as a more effective alternative.
- The truck dump is one particular site in the plant that requires effective dust control. This is a particularly difficult area to control and may require the combination of several dust control methods to obtain a satisfactory result.
- Vehicles often need to be washed before leaving the property to prevent fugitive materials falling from them onto the public roads.

Bulk Transportation

In General
- There are a number of systems used for transporting materials in bulk and for loading and unloading those systems. Material can be carried by ship, barge, train, and truck; material can be loaded and unloaded by belt, screw conveyor, pneumatic conveyor, or clamshell. The trend in the industry is higher "tons per hour" rates for loading and unloading, to reduce demurrage charges.
- Ship unloading is particularly challenging. On self-unloading ships, high angle conveyors that sandwich the material between two belts are often used. These typically operate at speeds greater than 900 fpm (4.5 m/sec). Equipment used in ship unloading systems must be able to withstand rugged conditions, including temperature extremes and exposure to salt water.
- Waterways (especially international waterways) and ports share use between commercial enterprises and recreational users. Dust and spillage over water in these areas will quickly draw complaints to regulatory agencies from other non-commercial users.
- Railcar unloading is a particularly difficult situation for effective dust control and can be a difficult material flow issue.

Material
- All types of material–from coal to chemicals, limestone to raw, and processed ores to grain and cement–are carried in bulk. It is important that the systems be designed to be flexible and work with a variety of materials. In many cases, different materials are loaded on the same conveyor, and contamination between cargos is an issue.

Conveying
- Belt tracking is critical to keep the cargo on the belt and to prevent spillage along the conveyor. The use of multi-pivot belt tracking devices is effective in controlling tracking problems.
- Access is often sacrificed on shipboard conveyors for space and weight considerations. This will increase the difficulty of maintaining conveyor accessories and therefore reduce the quality and frequency of service, leading in turn to excessive dust and spillage. Access is often difficult or impossible on boom discharges, and consideration must be given to portable work platforms or the ability to move the discharge into a location where service can be performed. Incorporating service access will greatly improve the effectiveness of accessories, as they will receive more frequent maintenance.
- The elimination of spillage is especially important (in loading applications), as the material is often considered tainted and unusable if it falls on the ground or into the water. In more and more jurisdictions, this material is considered hazardous waste.
- Spillage on trippers is a common problem, as they must reach minimum loading heights and carry various cargoes. Skirt walls the entire length of the incline and hanging deflectors help control rollback of material. Spillage trays are often placed under the conveyor in critical areas with provisions for constant flushing or easy cleaning.
- Belts on docks have to be designed with greater than normal edge distance for sealing to control spillage. Since these conveyors are often mounted on structures that must flex, they are prone to mistracking.

Belt Cleaning
- On high-speed conveyors and discharge pulleys that are difficult to reach, long lasting belt cleaners that automatically maintain contact cleaning pressure and angle are the best alternative.
- Washboxes have proven useful in making sure that the belts are very clean that travel over

bodies of water. Washboxes are also effective in reducing contamination when belts are used to handle several different materials. Fresh water must be used in washboxes and for washdown, or equipment will quickly become corroded.
- Air knives and vacuum systems are effective on very fine materials like alumina. These materials often exhibit a "static cling" tendency where the scrapings will flow right back onto the belt; in these cases, a vacuum pickup is needed.

Dust Management
- Dust collection is typically used, ranging from huge central systems to individual point source collectors.
- Conventional methods of dust and spillage control are often not sufficient. The use of washboxes and maintenance schedules as frequent as once every loading/unloading cycle are necessary.
- Special curved loading chutes are often used to centralize the materials and reduce dust generation.

Cement

In General
- This industry is truly global with new technology quickly adopted at a corporate level. There is a great deal of information sharing and focus on solving problems in material handling. The general trends in this industry directly reflect the global economy. Cement is readily shipped worldwide in large quantities.
- The cleaning and sealing of very fine, dry, abrasive material is a continuing challenge.
- The need for systems to handle waste-derived fuel, ranging from tires to diapers, will probably continue to grow.

Material
- Materials encountered in this industry run from large lumps in the quarry to high-temperature materials at the clinker cooler to fine, dry powders that risk fluidization in the packaging/shipping operation. Finished cement must be kept dry. A common problem is the elevated temperatures experienced when handling clinker.

Conveying
- In the quarry, applications require medium to heavy-duty ratings for conveyor equipment.
- Primary crusher discharge belts are often subject to huge impact forces that should be evaluated carefully to reduce belt damage, spillage, and leakage.
- On the cement processing side, belts are smaller and impact is not generally an issue. After crushing, light-duty ratings are generally sufficient.
- Raw and finished cement belts are excellent applications for air-supported conveyors to reduce spillage and contamination.
- Clinker is abrasive and often handled at high temperatures. This requires special belting and accessories designed for the elevated temperatures.

Belt Cleaning
- Cleaning clinker belts may require specialized high-temperature belt cleaning systems.
- When cleaning belts used for waste-derived fuel, single-blade pre-cleaners are preferred to reduce the chance of material collecting in the gaps between the blades.

Dust Management
- For dust control in the raw material stockpiles, water has been the suppression system of choice. The use of foam suppression is effective at the crusher and provides some residual effects.
- On the finish side, the addition of moisture is not allowed, so containment and collection are the only options.
- Air-supported conveyors can be effectively used in this industry for the control of dust.
- Because of the abrasive nature of the clinker and the very fine particle size of the finished product, leakage from chutes and skirt seals is a common problem. Extra attention to repairing and sealing holes in chutes will pay-off in dust control. Belt support and self-adjusting seals are useful in controlling dusting at transfer points.

See also the Aggregate Industry

Coal Mining (Underground)

In General
- Height restrictions are a major factor in this application. This will affect the style of conveyor structure, which in turn affects the accessory systems that can be used. Due to difficulties in movement and installation, chute work is minimal and impact cradles are rarely used underground.
- The low headroom and confined spaces make the transportation and installation of equipment underground more difficult. Modular designs are needed for many components.
- The trend is to use wider conveyors at higher speeds. Main conveyor lines will generally be vulcanized, but other belts may contain a large number of mechanical splices. There is significant use of used belting, as well as use of belts that are well past their true service life. This means the belts are in rough condition and, therefore, more difficult to clean, seal, and track.
- The belts that feed the main lines are designed to be extended. As the working face moves, additional sections (panels) of belting are spliced onto the belt. This means these belts incorporate multiple splices.
- Regulatory approvals (based on safety issues) are a factor in the selection of materials for components. In the United States, MSHA sets standards for conveyor belts and the materials in contact with the belt, such as cleaning and sealing systems. Outside the US, British and DIN Standards are widely accepted for materials used underground. In most markets outside of North America, aluminum cannot be used underground because of its low sparking threshold.
- Some countries require anti-static fire resistant materials, commonly called FRAS, be used for underground conveyor components in contact with the belt; other nations require only fire resistant ratings equivalent to MSHA requirements.
- The use of specialized contractors for accessory maintenance is common in this industry. This is due to the critical role belt cleaning and sealing play in preventing fires, explosions, and production outages.

Material
- Underground operations typically use lots of water to suppress dust. This solves one problem but creates others. The moisture will affect the properties of the material conveyed and the design of equipment, down to type of metal used.
- With modern mining methods, the lump size is fairly consistent. However, run-of-mine coal containing rock and clay will present handling problems.

Conveying
- Conveyors are designed to be moved with minimal effort, except for main lines and slope belts that carry material out of the tunnels. Belts are larger, thicker, and typically require heavy-duty cleaners and skirting systems. Self-adjusting skirting systems are useful.
- Transfers are often at 90 degrees or have large drop distances, creating heavy impact situations and adding difficulties in sealing. Plugging at transfers is common, and air cannons are useful in solving these problems.
- Spillage is difficult to control due to constantly varying loading conditions creating belt-tracking problems. The use of heavy-duty tracking devices is necessary because of the constant impact of mechanical splices.

Belt Cleaning
- The material handled in underground operations contains a great deal of water with the coal. This worsens carryback problems. As a result, many operations use three, four, or more cleaners on a belt.
- Many operations use multiple cleaner arrays with scavenger conveyors to carry the removed material back from the tertiary cleaners to the main material body.
- Because of the numerous splices, belt cleaners and mounts must be designed to handle the repeated impact. Urethane pre-cleaners with heavy blades for durability are preferred. A wide variety of secondary cleaners are used.

- The refuse belt is usually the most difficult to clean and produces the highest rate of blade wear. Aggressive cleaning with frequent maintenance is required in most cases.

Dust Management
- Water is used as a dust suppression agent at the working face.

Foundries/Metalcasting

In General
- As might be expected, high material temperatures and challenging service conditions are common.

Material
- Dusty/warm/moist materials can break-down urethane products.
- While foundry sand is not highly abrasive, its moisture content can lead to corrosion even in an abrasion-resistant plate.
- Sharp pieces of metal will occasionally slip through with the return sand.

Conveying
- Belt speeds are not high; they are typically in the range of 50 to 200 fpm (0.25 to 1 m/sec).
- Applications in foundries are typically light duty, with exception of reject/return systems where metal pieces the size of engine blocks will occasionally appear.
- Belt tracking is often affected by spillage of the sand, which tends to build up quickly on return rollers. Multi-pivot tracking devices can be used effectively.

Belt Cleaning
- Softer urethane blades may last longer in cleaning the round particles found in foundry sand from the relatively slow-moving belts.
- Brush cleaners may be effective in removing sand from worn belts.

Dust Management
- Containment and collection are the choice for dust control. Avoid adding moisture to molding sand.
- Foundry dust is often considered hazardous due to its silica content.

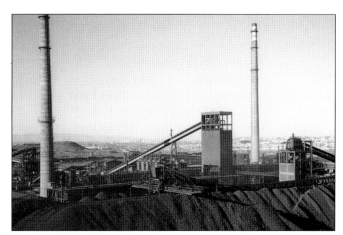

Metal Mining & Production

In General
- This industry is generally considered the "hard rock" mining segment because many of the minerals are contained in hard rock formations.
- Because metal prices fluctuate, budgets are frequently subject to change and projects are often put "on hold" or accelerated.
- Plants are typically in operation 24 hours per day, 7 days per week when metal prices are favorable. Outages are scheduled far in advance.

Material
- Ores are often mined by blasting, which results in large lump sizes. Discharge from primary crushers is typically 8-inch (200 mm) minus in size.
- Ores are generally highly abrasive material, which can shorten the life of the belt and other components.
- The process to make taconite pellets creates situations where high temperatures are common. Taconite dust can become embedded into the edge of the belt, becoming a "grinding wheel" that will saw off a guide roll from a belt training idler in a matter of weeks.
- Other ores such as nickel or bauxite are often found in clay-like formations that result in materials with very sticky, slick, and/or agglomerating characteristics.

Conveying
- Applications for conveyor equipment are typically on the heavy-duty end of the spectrum, featuring heavy loads and multiple splices on relatively short conveyors. Belt life is often so short that the belt is considered sacrificial, and so more aggressive cleaning and sealing equipment can be used.
- Overland conveyors are often used for transporting materials and waste. These conveyors are often difficult to access, and they may cross over sensitive areas such as highways or nature preserves.
- Belt widths of 72 inches (1800 mm) and speeds of over 1000 fpm (5 m/sec.) are common. The belts are often steel cord belts operating under high tension. The use of steel cables in the belts introduces new challenges to accessory equipment. It is common for damaged steel cords to protrude from the cover and "whip" belt cleaner blades and frames.
- Because belts are often loaded to capacity, spillage along the carrying run is common and the potential is high for large rocks to be bounced onto the return side of the belt which end up between the pulleys and the belt. Heavy-duty belt plows are used to prevent damage to the belt and pulleys.
- Abrasion is a significant problem, creating maintenance problems that reduce the effectiveness of accessories and the overall operation of the system. Chute liners and pulley lagging are typically high wear items. Bolt-on wear liners are commonly used, and the need for access to these bolts limits the skirt seal options.
- Belt tracking is a common problem on overland conveyors. Heavy-duty multi pivot tracking devices can be used in place of V-return rollers.
- Plugging of chutes is common. Air cannons are typically used to reduce this problem.
- Because of the high-capacity transfers and large lump sizes, impact is extremely high. Normal impact cradles may not be able to withstand the impact forces, and catenary impact systems may be required.

Belt Cleaning
- In steel production, after the bentonite clay is added, the material becomes very sticky. A metal-bladed pre-cleaner is useful for this application.
- In nickel and bauxite applications, the material is thixotropic in manner and difficult to clean from the belt. The use of water to keep belt cleaners free of buildup will enhance performance.
- Secondary cleaners usually have tungsten carbide blades to improve performance and extend life in the face of abrasive material. Heavy-duty cleaners with extra thick tungsten carbide blades are often used.
- Blade life is considerably shorter than other applications, and cleaners require frequent service to remain effective. Cleaners with constant tension for the life of the blade are necessary to keep service intervals to a reasonable period.
- Large lumps can ricochet in the chute and damage belt cleaners, either by direct impact or by becoming lodged in unusual places, disabling the tensioner or bending the cleaner frame. The use of extra-heavy-duty pre-cleaners is required. A system of two pre-cleaners often provides acceptable cleaning and is less susceptible to damage.
- Pulley cleaning is important on steel cord belts to prevent over tension and puncture of the belt from material buildup. Arm-and-blade secondary cleaners are often used "upside down" in this application in combination with a rock guard or deflector plate to remove the cleaned material from the belt.

Dust Management
- Dust suppression can be used not only in mining but also in pellet transportation. Collection is also common throughout the plant.

Power Generation (Coal-Fired)
In General
- In the utility industry, deregulation is driving down profitability. To improve reliability and

reduce costs, management is seeking to do more with less.
- Control of dust is a major issue in the industry now, particularly for plants that switched to low sulfur coal or that burn lignite.
- Dust emission regulations are impacting coal handling, from the railcar unloading station all the way through the material handling system to the bunkers above the boilers.
- All coal-handling operations need to be concerned with fire and explosion issues as well as possible methane gas accumulations and material "hot spots."

Material
- Day-to-day changes in material will affect conveyor performance. Due to conditions, the coal can range from very wet/muddy to very dry/powdery.
- Generally speaking, coal is a relatively low-abrasion material. The exceptions are raw coal or refuse belts as seen at mine-mouth generating stations.
- It is becoming more common to burn auxiliary fuels in combination with coal. These fuels include shredded tires and agricultural waste. The introduction of these materials into the coal is critical and if not well metered, will create spills, plugs, and other operating problems.
- Accumulations of fugitive material present serious fire/explosion potential from spontaneous combustion, and small events create high dust concentrations and possibly secondary explosions.

Conveying
- Conveyor belts are generally vulcanized and have a long service life.
- Conveyor widths are moderate, with widths from 36 to 54 inches (900 to 1400 m) common and speeds of 400 to 600 fpm (2 to 3 m/sec.) typical.
- Coal conveying is particularly well suited for air-supported conveyors.
- Belt tracking problems in coal handling can be solved with multi-pivot tracking devices. Standard pivot devices often over-steer on coal handling belts; that is why these belt trackers are often tied off to one side or another.
- Skirt sealing is important in coal handling. Coal conveyors are particularly well suited for belt support cradles and self-adjusting sealing.

Belt Cleaning
- The cleaning of coal-handling conveyors is usually rather straightforward and can be considered the typical application. A standard power plant belt cleaning system is a dual or triple system with a urethane pre-cleaner and one or two tungsten carbide-tipped secondary cleaners.
- Some coals contain clays that create difficult cleaning. This material tends to smear on the belt and accumulate as "corn flakes" under return idlers. The normal solution is to operate the belt cleaners at a higher cleaning pressure or to use a more aggressive cleaning angle.
- The use of water is beneficial in maintaining belt cleaner efficiency, but power plants often have ill-conceived edicts of "no water" due to the BTU penalty. The amount of water required to maintain belt cleaning efficiency is so small that it cannot be distinguished from other sources of water, such as dust suppression, rain, and even the water absorbed from high humidity.

Dust Management
- Low-sulfur coal burns cleaner, but is typically more friable. As plants move to the cleaner-burning coal, they will need to upgrade their existing dust collection systems (baghouses).
- Water-only dust suppression is not cost-effective, as it reduces the BTU output of the coal. Chemical suppression is the choice of most plants, as the reduced moisture levels minimize BTU penalty.
- Rotary railcar dumpers for coal unit trains create large problems with dust generation. Foam or surfactant suppression offers benefits including a residual effect that stays with the coal as it goes into the stockpile.

Pulp & Paper
In General
- Conveyors are used in moving logs to the chipper, carrying chips to the digester, and bark as fuel to the plant's power generation

system. As the chips are a precious raw material, and the bark is only a waste by-product, the chip handling system is more carefully maintained. The bark handling system, due to its cargo of oddly sized and stringy material, is generally more of a mess.
- Plastic can contaminate the pulp made from the chips. Consequently, there is a concern with the use of plastics on process and material handling equipment. Some plants will forbid the use of urethanes as belt cleaners, for example. But this is typically a plant-level decision; there is no set rule in the industry or generally even at the corporate level.

Material
- The materials conveyed in pulp and paper operations pose some problems. The fibers tend to interlock easily, leading to buildup of stringy material in chutes and equipment and to plugging of chutes in the wood yard. Wood chips contain resins that build up on the belt and rollers, and is very difficult to remove.

Conveying
- In general, the duty rating for applications in this industry is light. The exception is at the debarking drum, where the tree-length or cut logs are discharged onto conveyors to be moved to the chipper.
- Sticks and chips tend to catch under skirting and can plug-up the loading zone or cause spillage. Careful attention to loading chute design with tapered loading chutes and gradually increasing liner height helps prevent this. Low-pressure, self-adjusting skirt seals tend to be more "self cleaning" than fixed seal systems.
- The use of air cannons needs to be reviewed carefully, as wood chips and bark often require much larger air cannons than a typical application, due to the porosity of the bulk solid.

Belt Cleaning
- The presence of wood resin or pitch on the belt creates cleaning problems. This is difficult to remove by itself, and it causes other materials–bark strands, chips, or fines–to stick as well, complicating the cleaning process. Thin tungsten carbide-tipped secondaries operated at higher than normal pressures are sometimes needed to solve this problem. Also, cycling the belt with no load and allowing the cleaner to remove the buildup after each production cycle helps prevent the material from setting up.
- The discharge of the digester feed conveyor is a difficult application for urethane pre-cleaners, as the digester chemicals soften the elastomer blade and so shorten life.
- Cleated belts are often used to move the chips or bark up inclines. Brush cleaners and chevron belt cleaners are required to clean these belts.
- On bark-handling conveyors, "arm and blade" secondary cleaners tend to accumulate buildups of stringy material that interfere with performance. More success will be had using in-line cleaners without arms or single-bladed cleaners.
- Special blades may be required for white paper production to prevent contamination of the paper with colored particles from belt cleaner wear.

Dust Management
- Chemical suppression is used throughout the system, although rate of application may be restricted depending on position in the process.
- Dust collection systems are common; in most cases, these are large central collectors.

Surface Mining
In General
- The large quantities of material handled for these operations leads to the use of oversize equipment–from bucketwheel excavators to haul trucks, to wide, fast belt conveying systems.

Material
- Materials can range from lignite and low rank coal to ores for base and precious metals. In addition, large quantities of overburden must be moved. This overburden can change greatly in material characteristics as different layers of strata are removed on the way down to the ore level.

Conveying
- Wide, high-speed, high-capacity conveyors are the rule rather than the exception. For example, German lignite operations use conveyors with belt widths up to 124 inches, (3200 mm) operating at speeds up to 2100 fpm (10.5 m/sec). Typically these operations are pushing equipment suppliers for bigger, faster, higher-tonnage systems.
- There are often extreme levels of impact in conveyor loading zones from uncrushed, run-of-mine materials. Loading zones should be designed for these forces with impact idlers and impact cradles. To handle this impact, many operations incorporate catenary idlers, creating difficulties in the sealing of load zones.
- Changing material characteristics–from different layers of overburden for example–may allow accumulations that can choke and clog chutes. The installation of air cannons on transfer chutes may be useful.

Belt Cleaning
- High belt speed and material velocity lead to high frictional temperature and high vibration levels. Belt cleaners must be engineered to withstand these conditions. Pre-cleaners with a high volume of urethane are often used to extend service life and dissipate heat.
- High conveyor operating speeds may not allow higher pressure secondary cleaners to be installed. However, larger head pulleys may have enough room for two lower-pressure pre-cleaners below the material trajectory.
- The continuous production schedules and high tonnage conveyed will require systems engineered for long life and low maintenance. In some operations, this may allow the use of "maintain while in service" systems; i.e., belt cleaners installed on mandrel-style or telescoping mounts that allow the blades to be changed while the conveyor runs. The adoption of this maintenance practice will require approval of the appropriate regulatory agency and internal safety department.
- Return belt cleaning is important because of the potentially large size or sticky nature of material. These materials become trapped between the belt and bend pulleys and can damage the belt either by puncturing it or increasing the tension. The buildup of materials can also quickly cause mistracking. Return-side belt cleaning devices must be designed for high impact and to prevent entrapment of materials in the suspension systems. Pulley cleaners are often applied in addition to belt plows.
- Cleaning devices for the inside of the belt must be designed for high impact and to prevent entrapment of materials in the suspension systems. Pulley cleaners are often applied in addition to belt plows.

Dust Management
- Spray-applied water is the typical dust suppression method. However, high rates of water application will increase problems with carryback or screen blinding. Surfactant or foam suppression can be considered as an alternative.
- The truck dump leading to a crusher will generally require dust control systems.

See also Coal Mining, Metal Mining & Production

Chapter 17

Safety

Belt conveyors and their transfer points can be dangerous. By their very nature, they feature many "pinch" points and speeding objects. They apply large amounts of mechanical energy to what is basically a loaded rubber band. Operations and maintenance personnel must always be aware of the power of a conveyor, and they must maintain a healthy respect for its potential to injure or kill an untrained or unwary individual.

However, it should be noted that conveyors can be one of the safest ways–if not *the* safest–to move the large quantities of material required for large-scale industrial processes. Other forms of bulk haulage–from trucks to trains to ships–all carry their own risks and safety concerns, and may in fact be more unsafe than belt conveyors. As CEMA noted in *Belt Conveyors for Bulk Materials,* few personnel are required for operation of conveyors and they are exposed to fewer hazards than other means for transportation.

While it must be recognized that accidents can happen, it must also be observed that they can be prevented. Conveyor safety begins with a design that avoids foreseeable hazards. Plant management must require equipment that is safe and easy to maintain. Installations must be designed with free entrance and exit from conveyor transfer areas. Vendors should provide information and training on the safe use of their products. Management must insist upon good housekeeping practices in terms of the cleanup of fugitive material, but also in terms of the disposal of discarded machine elements and packaging materials,

The establishment and maintenance of safe practices in the design, construction, operation, and maintenance of conveyors and conveyor transfer points will aid greatly in the prevention of accidents. Guidelines for safety in conveyor design and operation have been spelled out by CEMA in *Safety Standards for Conveyors and Related Equipment* and adopted by the American National Standards Institute in ANSI B20.1-1976. These works and/or their international equivalents should be consulted as a guide for the design and construction of any belt conveyor system.

General Safety Guidelines

There are certain safety procedures that should be observed regardless of the design of the conveyor or the circumstances of operation. They include:

1. Lock out/tag out all energy sources to conveyor belt, conveyor accessories, and associated process equipment before beginning

> "While it must be recognized that accidents can happen, it must also be observed that they can be prevented."

any work–whether it is construction, installation, maintenance, or inspection–in the area. Use a lock with one key for each piece of equipment. Only the person doing the work should hold the key to the lock.

2. Employees should be properly trained on the material handling system (including emergency warning signals and emergency stop controls) before they are allowed to work in the area.

3. The conveyor safety program should be one portion of a comprehensive and formal safety program within the facility in which all employees participate.

4. A formal inspection and maintenance schedule should be developed and followed for the material handling system. This program should include emergency switches, lights, horns, wiring, and warning labels, as well as moving components and process equipment.

5. Employees and visitors should learn and observe all applicable company, local, state, and federal safety rules. It is recommended that a formal testing program covering the safety rules be conducted annually.

6. The recommended operating speed and capacity for the conveyor and chute should not be exceeded.

7. All tools and work materials should be removed from the belt and chute before restarting a conveyor. A safety "walk around" is recommended prior to conveyor operation.

8. All emergency controls should be close to the system, easy to access, and free of debris and obstructions.

9. A hard hat, safety glasses, and steel-toe shoes should be worn when in the area of the conveyor. Loose or bulky clothing should not be allowed.

10. Never poke, prod, or reach into a material handling system while it is in operation.

11. Never allow personnel to sit on, cross over, or ride on an operating conveyor.

The above guidelines are not intended to replace the more detailed safety guidelines published in ASME Standards B20.1 and B15.1. Consult those references, as well as the safety instructions provided by the manufacturers of specific systems.

Lockout/Tagout

A crucial procedure in the conveyor safety program is lockout/tagout. In the United States, this is an Occupational Safety and Health Administration (OSHA) requirement; the Mine Safety and Health Administration (MSHA) has adapted a similar version of this rule. The lockout/tagout rule requires that power to the conveyor system (and any accessory equipment) be shut down, locked, and tagged by the person performing work on the conveyor system. **(Figure 17.1)** Only the person who locked out the system can unlock it. This prevents someone from starting the conveyor belt unknowingly while someone else is working on it.

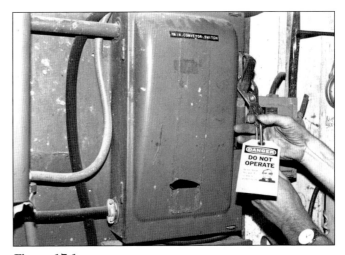

Figure 17.1
Safety rules require that the power supply to the conveyor be shutdown, locked out, and tagged by personnel working on the conveyor.

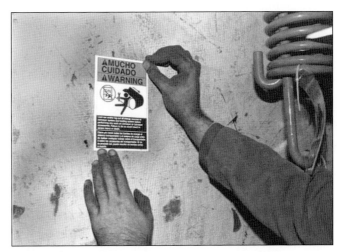

Figure 17.2
Safety stickers should be affixed at hazardous points on conveyors and other plant equipment.

Conveyors Can Be Deadly

As he was walking the length of the 150-foot-long (45-meter-long) conveyor belt that ran from the crusher to the storage bin, he noticed that the take-up pulley was caked with gunk again. He'd worked in the mill for 10 years now, and for all that time the scraper on the head pulley that was supposed to keep the belt clean seemed to constantly need adjusting. Oh well. He'd get to that in a minute. First to clean the take-up pulley.

The crusher operator was away from his post and both the crusher and the conveyor were shut down. He walked back to the control box, turned on the conveyor, and then got a front-end loader and drove it back to the take-up pulley. He raised the bucket of the loader high enough so that he could stand in it and reach the pulley. Grabbing a tool–an ordinary garden hoe with a handle cut down to 16 inches (40 cm)–he climbed in the bucket and began to scrape the gunk off the pulley. He'd done it this way several times before and it was so much faster when the pulley was turning. And the belt was only moving 450 feet per minute–that's only about 5 miles (8 km) per hour.

Before he even knew what was happening, something seemed to grab first the tool and then his arm and yanked him up out of the bucket. The pain of his arm being crushed lasted only until his shoulder and neck smashed into the conveyor structure. He died instantly. He was 37.

This story is not fiction. It is also not the only one that could have been selected. Conveyors kill all too frequently–or amputate, cut, or crush.

Because conveyors are so common, it is easy to take them for granted and become complacent. And at about 5 miles (8 km) an hour, they don't seem to be moving that fast. It's easy to believe that the victim in this story could have simply dropped the tool when it was caught. Not so.

Five miles an hour (eight km/h) works out to about 7 feet 4 inches (2.2 meters) per second. At that speed, if your reaction time is three-fourths of a second, your hand would travel 5.5 feet (1.6 meters) before you let go of the tool. Even if your reaction time is a fast three-tenths of a second, your hand would travel over 26 inches (660 millimeters). That's far enough to become entangled if you're using a short tool. And if it's a loose piece of clothing that gets caught, then it doesn't matter what your reaction time is. It's already too late.

*Reprinted with permission from the **Safety Reminder** newsletter, published by Ontario Natural Resources Safety Association, P.O. Box 2040, 690 McKeown Avenue, North Bay, Ontario, P1B 9P1. Telephone (705) 474-SAFE.*

Safety Stickers

Safety stickers and warning labels should be affixed at pinch points, service access doors, and other hazardous areas on conveyor equipment. **(Figure 17.2)** These should be kept clean and legible, and reapplied appropriately to suit changes in equipment or procedures.

Safety stickers and signs are available from reputable manufacturers of conveyors and related equipment, as well as safety supply houses.

Emergency Stop Switches

To protect personnel working near the belt, the conveyor should be equipped with "pull rope" emergency stop switches. These safety switches should be conveniently mounted along conveyor catwalks and right-of-ways and run the full length of both sides of the conveyor. The switches should be wired into the belt's power system so that in an emergency, a pull of the rope interrupts the conveyor drive power.

In 1995, the Mine Safety and Health Administration alerted mine operators about potential failures of emergency-stop pull-cord systems along conveyors. After tests of over 1100 systems, MSHA noted a failure rate of two percent. MSHA attributed these problems to several factors:
- Spillage around the switch that prevents deactivation of the conveyor.
- Broken pull cords or excessive slack in cords.
- Frozen pivot bearings where the switch shaft enters the enclosure.
- Failure of electrical switches inside the enclosure.
- Incorrect wiring of switch or control circuits.

The solution to this problem is, of course, proper service attention and testing drills similar to school fire-drills, when the operation of the conveyor safety equipment can be checked.

Warning horns or bells should be located at each transfer station to indicate when the belt is about to begin moving.

Guards On Equipment

It is important that pinch points on rotating equipment like head pulleys, and on equipment that allows sudden movements, like gravity take-ups, be shielded to prevent accidental or unwise encroachments by employees.

At the same time, it must be remembered that service access must be provided to the various pieces of equipment. The physical barriers installed to shield this equipment must be carefully designed, or they will interfere with maintenance efficiency. After the appropriate lockout procedure is carried-out, the guards must be removable in an efficient manner.

All guards, safety devices, and warning signs should be kept in proper positions and in good working order. After service is completed, it is important that guards be returned to position prior to restarting the conveyor.

Personal Responsibility

Safety is not just the responsibility of a plant's safety department. Rather it is the responsibility of each worker to work in a safe manner, assuring safety for oneself and for one's coworkers.

The first step is to take personal responsibility for one's own safety. The measures that only the individual can make certain of include the use of personal safety equipment, including dust masks and respirators, hearing protection, hard hats, and steel-toed shoes. Appropriate clothing should be worn. If it is too loose, it could get caught in the conveyor–and long hair and any jewelry should be secured.

Preparations for this personal responsibility include attention to safety practices as trained by the company safety department and required in the on-site safety program, good standards of housekeeping to provide a clean and safe work area, and a thorough review of equipment manuals to learn safe operating and maintenance procedures.

In many ways, plant safety is like plant cleanliness; they are both matters of attitude. Plant management can set a tone for the overall operation. However, it is the response of each individual that will have the greatest impact on safety.

Chapter 18

Access

"The providing of proper access to equipment is a problem that will not be solved by magic."

Like all mechanical systems, conveyors and transfer points require regular maintenance activities for continued operating efficiency. To facilitate this maintenance, conveyors, chutes, and loading zones must be designed to accommodate service.

Too often systems are not designed in ways that will both minimize maintenance requirements and facilitate service when it is required. The most basic of these provisions is to include in the design ways to access the components that will need service attention. In an article on improving maintenance practices, G.C. Simpson noted, "It has been shown in industries as diverse as mining and nuclear power generation that simply gaining access to the machine can often account for over 30 percent of the total maintenance time."

More than just lengthening the time needed to perform maintenance (and thus extending the "outage" time that the equipment must be shut down), difficult access may mean that important maintenance chores do not get performed. If a system is difficult to maintain–either because the maintenance activity is difficult or because the access is awkward or time-consuming–the service work may be postponed or not performed at all. Or it may be done in a cursory, sloppy, path-of-least-resistance fashion. Either way means greater risk of component failure and the resultant loss of production.

The providing of proper access to equipment is a problem that will not be solved by magic. Access requirements must be carefully considered during the design of a conveyor system.

Conveyor Spacing

Many times, in order to save costs, conveyor equipment is placed in small galleries or enclosures. **(Figure 18.1)** One side of the conveyor is typically butted against a wall, an adjacent conveyor, or other equipment. It is extremely difficult to service this type of installation. If the conveyor is installed flush against a wall, vessel, or other structure, basic service requirements such as bearing lubrication or idler replacement become major operations requiring extended production outages.

A better system would be to allow a service space between any conveyor's outermost structural member and any permanent obstruction, such as a wall or another conveyor. This space must provide adequate room for service personnel and their equipment. It is necessary to provide at least 24 inches (600 mm) or one-third the belt's width, whichever is greater, on both sides of the conveyor. **(Figure 18.2)** This will facilitate the variety of maintenance chores that seem unimportant during the initial design of the conveyor, but that turn into critical necessities after the conveyor is operational.

Generally, there must be sufficient room to allow access to all sections of the conveyor system and, in particular, both sides of the conveyor. Failure to provide access to both sides of the conveyor is the most common deficiency in making conveyors operable and maintainable.

Walkways and Work Space

Proper access requires the provision of fenced walkways and platforms with ample headroom. These should provide a firm path alongside the conveyor and around head and tail pulleys with access to all lubrication, observation, and other maintenance points. Walkways should be a minimum of 24 inches (600 mm) wide for passage, and 36 inches (900 mm) wide in areas where service work must be performed. Where a person must kneel to perform service or inspection, a "head room" of at least 45 inches (1200 mm) should be provided. Areas where frequent service or cleaning is required should have solid flooring.

When conveyors run parallel to each other, the space between them should be a minimum of 36 inches (1000 mm) or width of the belt, whichever is greater, to allow for belt repair and removal of idlers.

Another common deficiency in conveyor design is failure to allow adequate space for cleanup. A study of conveyor-related accidents in mining showed that one-third of all accidents occurred to workers trying to clean up under or around the carrying run of the conveyor. Areas that require frequent cleanup should allow for mechanical cleaning, such as the use of a skid-steer loader or vacuum truck under the conveyor. If this is not practical, a minimum clearance of 24 inches (600 mm) between the bottom of the return rollers and the floor is suggested. **(Figure 18.3)**

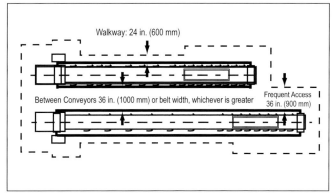

Figure 18.2

Proper spacing between conveyors will allow efficient service.

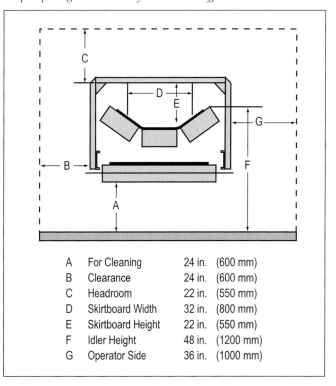

A	For Cleaning	24 in. (600 mm)
B	Clearance	24 in. (600 mm)
C	Headroom	22 in. (550 mm)
D	Skirtboard Width	32 in. (800 mm)
E	Skirtboard Height	22 in. (550 mm)
F	Idler Height	48 in. (1200 mm)
G	Operator Side	36 in. (1000 mm)

Figure 18.1

Conveyors are too often installed against a wall, an adjacent conveyor, or other equipment.

Figure 18.3

Cleaning around the conveyor will require sufficient space.

Access platforms, ladders, crossovers, and other equipment necessary to reach the far side of the conveyor should be included in the system's design and specifications.

A surprising number of accidents happen to personnel who are riding on the belt or crossing over or under the conveyor. Crossover platforms should be provided, and riding on the belt should be prohibited.

The Encroachment of Other Systems

It is not uncommon to see a conveyor or transfer point captured in a web of electrical conduit, dust suppression piping, control panels, or sprinkler systems. **(Figure 18.4)** Any attempt to reach the components of the transfer point must first bypass the "thicket" surrounding the transfer point. The interruption of these other systems results in a variety of other complications to the plant's operations.

To control the growth of these auxiliary systems around the conveyor and transfer point, the designer must specify the equipment to which access is necessary. By providing specific areas in the conveyor's plans for the installation of control panels, gate actuators, and other equipment, many unnecessary obstacles can be avoided.

Observation Requirements

Very often, flow problems within chutes can be more easily solved if the material path can be observed. The actual path of material within a chute cannot always be predicted, so observation is necessary to allow adjustment of diverters, gates, and grizzly bars. Many transfer chutes have only one inspection door. This is usually installed near the head pulley, and does not permit a view of the actual material path in the lower chute and skirted area where problems often occur.

The chute should incorporate observation apertures with easy-opening covers located away from the flowing material. These openings should allow safe observation of both the material flow and component wear at critical areas of the installation. The openings should be limited in size and/or protected with fixed bars or screens to prevent personnel from reaching in and material from flying out.

Consideration must also be given to providing ample lighting at access points. This may simply require sufficient overhead lighting or may necessitate the availability of high power spotlights or strobes that can be aimed into the chute to observe material movement.

Chute Access

It is important that any enclosed chutework be provided with adequate access. Not only must openings be installed–both inspection windows and worker entry ports–but paths must be left clear for workers to efficiently reach these openings.

At least two inspection or access doors should be provided in a transfer chute: one in the head chute and one in the loading zone. If the chute has more than a ten-foot drop, another door should be located at the approximate midpoint but out of the material trajectory. In situations where a drop of 20 feet or more occurs, a door every ten feet is recommended.

It is important to have access from both sides of a chute. Doors should be located in pairs, opposite each other, to allow for the insertion of ladders or planks for scaffolding. This arrangement will simplify any maintenance required inside the chute such as liner replacement, the freeing of blockages, and the removal of tramp materials.

It is similarly important to provide attachment points in or adjacent to chutes and access points. These allow the use of devices like bosun's chairs or come-alongs to improve maintenance procedures. If positioned within the chute, these attachment points should be well out of the material trajectory to avoid the abrasive wear.

Figure 18.4
Insufficient space will make access for required service difficult.

Figure 18.5
Access doors should close tightly to keep dust inside the enclosure.

Figure 18.6
Flexible rubber "snap-on" doors offer easy operation in locations with limited clearance.

The consideration of service is particularly important on chutes too small to allow personnel to work inside. Fabricating these chutes in sections for easy disassembly is one approach. Another option would be the flange bolting of the non-wearing side of the chute to allow the opening or removal of the entire side panel.

Door Construction

Inspection doors should be side opening and sized to view easily and safely the components inside the structure. Doors must be installed on the non-wearing sides of the chute–the sides away from the flow of impacting or abrasive material.

Doors should be designed for easy operation in tight clearances, with corrosion-resistant hinges and latching systems. It is important that all ports be dust-tight through use of a securely sealing door. Hinged metal doors with easy-opening latches are now available to provide access. **(Figure 18.5)**

Flexible rubber "snap-on" doors provide a dust-tight closure, while allowing simple, no-tool opening and closing even in locations with limited clearances. **(Figure 18.6)**

Door sizes should be large enough to provide the required access. If observation or service requirement is limited to systems such as belt cleaners, a 9-by-12 inch (225-by-300 mm) or 12-by-14 inch (300-by-350 mm) door is usually sufficient. If service to major components such as chute liners will be necessary or if personnel will need to use the door as an entry to the chute, then door sizes of 18 by 24 inches (450 by 600 mm), or 24 by 24 inches (600 by 600 mm), or larger will be necessary.

Remember to heed all appropriate shutdown and lockout programs when opening access doors for observation or entry.

It is important that access doors and covers be closed following use to avoid the escape of material and the risk of injury to unsuspecting personnel. It makes sense to add places where safety harnesses can be attached in line with access doors.

Chapter 19

Maintenance

The maintenance function in any operation is charged with keeping systems running at maximum productivity. System availability must be kept at maximum (or optimum), while the actual maintenance workload is reduced to the minimum required to ensure the efficiency and safety of the system.

No matter how well they are engineered and constructed, conveyors and their component systems will require timely maintenance to keep performance at design levels.

Although one seized idler may not appear important, it must be recognized that under the strain of a high-speed belt handling abrasive material, the idler's shell can soon wear through, becoming a knife edge that could severely damage an expensive belt. Well-trained personnel should be able to detect impending failure in such a case and correct the malfunction before any damage could occur. There are a number of other similar, seemingly minute problems that, under the constant stress of high-speed movement of belt and material, can quickly lead to costly damage. These problems, from the gradual accumulation of fines, to the chattering of belt cleaner blades, add expense in both equipment costs and unnecessary downtime.

The Cost of Unscheduled Outages

Emergency shutdowns are very expensive. One author listed the cost for service at from three to seven times more to do work durning an emergency shutdown than during scheduled downtime. Take, for example, the case where a long belt feeds a shorter one and they are stopped together. The longer conveyor takes a longer period to decelerate and may dump many tons of material into the transfer point. Significant damage can be caused, and much corrective work may be required to return overflow spillage to the system. Any means, therefore, of preventing such emergency shutdowns is to be highly valued.

In conveyor operations where downtime can easily cost from $1,000 to $5,000 per hour or more, the price of even the shortest unscheduled outage is prohibitive. One conveyor industry source chalks up the downtime as costing the average mine anywhere from $200 to $1,000 per minute. The downtime rate for longwall coal mines runs at $500 per minute or $30,000 per hour. Obviously, maintenance to prevent unscheduled downtime is a critical factor in the operation's overall profitability.

Depending on its size, a one-percent difference in on-line availability of a coal-fired power plant could be worth $1 Million to $2 Million US in

> "No matter how well they are engineered and constructed, conveyors and their component systems will require timely maintenance..."

annual revenue. Now, a conveyor outage or failure that costs even one-tenth of one percent in generating availability is a significant cost.

Preventive maintenance helps lower the total cost of operating a belt conveyor by protecting the expensive components and minimizing the risk of a catastrophic belt or system failure.

Ironically, plants that try for false economies in maintenance and cleaning usually end up paying more over the long term in environmental and cash terms to cope with cleaning and fugitive material control in their plant. The key to minimizing these unscheduled downtimes is efficient maintenance.

Ergonomics and Maintenance

Wherever personnel are involved in a system, whether as designers, operators, maintenance staff, or management, human performance and reliability will greatly influence the overall effectiveness and efficiency of that system. Ergonomics (or human-factor engineering, as it is sometimes known) is concerned with optimizing the relationship between the worker and the work. For a mechanical system to deliver its full potential, sufficient thought and commitment must be given to optimizing the human role in that system. Ergonomics now provides the means to complement mechanical reliability with human reliability.

While it is true that improved equipment reliability is the best route to reduce maintenance cost, the benefits of improved maintainability should not be overlooked. For example, achieving equipment reliability targets is critically dependent upon routine maintenance such as lubrication. In order to enhance human reliability, it is necessary to understand those situations that degrade human performance and increase the probability of mis-, mal-, or nonperformance. Armed with this information, it is then possible to influence the design of equipment to improve the reliability of operators and maintenance personnel and, thereby upgrade the overall efficiency and safety of the operation.

Designing for Maintenance

Maintenance management begins when planning the conveyor system. Too often equipment is not designed in ways that will both minimize maintenance and facilitate that maintenance when it is required. Designers rarely consider ease of maintenance when developing systems. In general, they do not like to think about maintenance; they see their systems as shiny and new, long-lived and enduring. In the designer's view, maintenance is a sign that something is wrong with the equipment, something that may be the fault of the design. As a consequence, the issue of maintainability is commonly ignored.

In the face of actual industrial conditions–with the continuous wear of day-to-day operations mingled with the demands of changing material conditions and amounts–system service becomes a very real requirement. The problem is magnified if the designers did not make adequate provisions for necessary service from the onset. For many plants, a common assumption is, "If it is difficult, time-consuming, or potentially dangerous, then it is inevitable that shortcuts will be taken." In the case of maintenance activities, this means that if a system is difficult to maintain, necessary service will probably not be performed. Or, if the work is performed, it will be done in a cursory, path-of-least-resistance fashion. Either way means greater risk of component failure and the resulting loss of productivity.

In many cases, the engineering process allows, or even promotes the development of maintenance problems. These problems include such built-in difficulties as awkward spacing and inaccessible positions, "permanent" fasteners, and other nonrepairable systems. In other cases, the designers have been careful to think of the larger service projects but neglected the routine chores. As an example, many conveyors include the provision of a frame or a way to lift out and replace the head pulley–an event that happens every three to five years. But there are no accommodations included for regular service like idler lubrication or belt cleaner maintenance.

There are ways available to solve these problems including adequate walkways and platforms, improved access, "slide-in/slide-out" track-mounted components, and "hammer" adjustability. It is this "designed-in" ease of service that simplifies and encourages routine maintenance to actually occur. The key is the consideration early in the design stage of the importance of maintenance.

Some equipment has been designed to allow some maintenance procedures to be performed while the belt is running. For example, belt cleaning systems have been developed that incorporate "telescoping" sections so the cleaner can be removed from cleaning position to outside the conveyor enclosure while the belt is running. Regulatory agencies and in-house safety departments must carefully review the use of these capabilities before it is applied in any specific operation. Any personnel undertaking this procedure must be thoroughly trained.

As maintenance crews are reduced and conveyor sizes increase, it becomes more important to supply the auxiliary equipment that will simplify maintenance. Well-placed access to utilities such as water, electricity, and compressed air will greatly speed maintenance work. Installing overhead hoists or jib cranes as part of the plant design, rather than as an after-thought will save maintenance time and improve safety.

Figure 19.1
Misalignment switches will shut off power to the conveyor when the belt wanders.

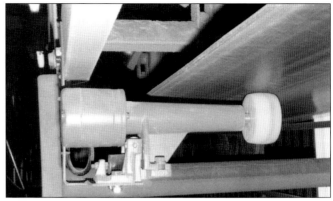

Figure 19.2
Belt speed monitors guard against conveyor overspeed or under speed.

Inspections

A non-programmed, general, day-to-day observation of conveyors by their operators can be invaluable in extending belt life and improving performance. Routine maintenance inspections can keep minor, easily-corrected problems from turning into major and costly headaches.

It is difficult to conduct a useful belt inspection when the belt is operating at speeds over 250 feet per minute (1 meter per second). Gross rips and other injuries may be spotted, but smaller defects will escape detection at these speeds. One solution to this problem is walking the full length of a stopped belt, checking it section by section.

In some of the best organizations, maintenance workers routinely "walk" the conveyor system, looking for indications of potential trouble. Irregularities such as gouges, edge fraying, punctures, and mistracking can be spotted and corrected promptly.

An alternative would be the use of a slow-speed inspection drive. Some conveyors have a "creep" drive (20 to 50 feet per minute) as part of their original design that will allow a slow-speed inspection. If not, a simple supplementary drive will permit visual observation of most belt injuries.

In addition, there are a number of controls and monitors that can detect problems with the conveyor and the belt.

Conveyor Control Systems

There are a number of devices that can be installed to help ensure the belt itself will be kept safe from accidental damage. These devices might include:

Misalignment Switches: If the belt tracks too far in any direction, these switches, installed at the limit of safe travel, are knocked over, cutting the power to the conveyor drive. **(Figure 19.1)** These "wander switches" should be installed at any point where the belt enters an enclosure or where a mistracking belt is within reach of structural steel or obstructions. Most commonly, they are installed on both sides of the belt at the head and tail of the conveyor.

Plugged Chute Switches: When the material "bridges" across the chute mouth, the material will buildup in the hopper above the discharge point. If this buildup goes too high, it may endanger the discharging belt above. Typically controlled by a paddle or an electric eye, plugged chute switches stop the feeder conveyor, halting the flow of material into the chute.

Belt Speed Monitors: Conveyors that run too fast or too slow can create problems that can damage the belt. Speed controls **(Figure 19.2)** can sound an alarm or shut down a system when belt speed deviates too far from the preset range. These may be used to check slippage, to prevent overloading of a belt that is suddenly not moving as fast as required, or to prevent overspeed, such as a runaway belt on a down grade. They may also be used to ensure proper sequencing of multiple conveyor installations or as alarms to prevent the sudden reversal–due to slippage or a power outage–of an inclined belt.

Conveyor Backstops: These are mechanical devices that allow a conveyor or drive shaft to rotate only in one direction. Automatic backstops should be installed on the drive pulley shaft of all conveyors or elevators that could reverse under loaded conditions.

Magnetic Devices: Stray metal objects of all types–ranging from crate strapping, to bucket loader teeth, to plate steel that has fallen from chute linings–can create a problem for many operations. If it is carried with the load on the belt, this "tramp iron" can wedge itself into a position where it can gouge or rip the belt. Magnetic head pulleys, which hold the iron on the pulley past the material discharge point and then drop it into a separate pile or chute, are commonly used to safeguard the belt. Another system uses a magnet suspended above the belt to pull the iron out of the load.

Rip Detectors: Large metallic objects in the load or pinned against the chutework can tear a belt lengthwise. If not detected, this slit can ruin the full length of a belt. Electromagnetic systems designed to sense any cut in the belt are now available to halt conveyor operation before belt damage is irreparable.

Belt Condition Sensors: These systems represent the latest electronic technology combining speed monitoring, tracking alarms, and rip detectors in one system. They incorporate a closed circuit sensor embedded in the belting during its manufacture. As the belt travels, these sensor loops pass over electromagnetic detector heads. Each detector head generates an output pulse that must be transmitted through the sensor loop to the input head. These systems can analyze the condition of the belt, including damage to cover and carcass and splice workmanship.

A rip in the belt will cause a break in the sensor loop so no pulse transmission can occur. The lack of a pulse immediately alerts the control unit to shut down the conveyor before further damage can occur. Similarly, lateral movement (belt wander) or belt slippage that results in a failure of the sensor loop to pass over a detector within a prescribed time will also cause a no-signal alert and shutdown. This system can even monitor its own operation, and reports on any malfunctions without affecting conveyor operations.

Electrical Interlocks: Each conveyor should incorporate electrical safety devices to protect both components and personnel. All electrical devices should be interlocked electrically so that when one conveyor "trips," the downstream conveyors and the upstream feeding devices also shut down. No conveyor can be started until the "tripped" safety device is checked and put back into service.

Fire Detection and Prevention Systems: Conveyor idlers can occasionally jam or lock due to lack of lubrication, material buildup, or mechanical breakdown. Once locked, the idler's bearings can overheat to a point where they can ignite flammable materials (like coal fines) that have accumulated around the roller. Similarly, friction from belt slippage can lead to overheating and fire. These fires can then quickly spread to flammable material on the belt and the belt itself. Many plants install complete conveyor fire protection systems including both heat detection sensors and a water spray system to extinguish any fire discovered. While these systems are not cheap, the costs for installing and maintaining these systems are substantially lower than what it would cost to repair or replace equipment damaged by fire or explosion. Fire detection systems based on infrared, ultraviolet, or fiber-optic principles are available. Once the bearing or belt overheats, it warms the air around it, activating the alarm, generally tripping the conveyor drive motor, and/or actuating a localized flooding system for the affected portion of the conveyor.

Closed-Circuit Television: Video cameras can be installed to observe the points where it is most likely for damage to occur and transmit images back to a central control room. An operator stationed in the control room can observe the monitors, watching for dangerous or damaging conditions.

There are many systems designed to provide safe conditions for both personnel and equipment. But the best guarantee of this crucial ingredient for any operation is the healthy respect of engineers, operations personnel, and maintenance staff for the power of the conveyor and the potential risks and rewards of its operation.

Maintenance Personnel

Some studies show that actual "wrench time"–the time that maintenance workers are actually performing maintenance work–is startlingly low. A recent study of the cement industry showed that "time on tools" for maintenance personnel was 15%. Measures to increase this time–from improved maintenance planning and to simplified workorder procedures would be well worthwhile.

It is important that belt conveyor maintenance be performed only by competent, well-trained personnel who are provided with proper test equipment and tools. For reasons of both safety and efficiency, the conveyor maintenance crew should be skilled, veteran employees empowered with the authority to shut down a conveyor to make a minor repair that will prevent a major outage or equipment expense.

The maintenance management system–computerized or not–cannot inspect the skirtboard seals or adjust a belt cleaner. These jobs are performed by maintenance workers. The system merely administers the work orders and manages information, so the maintenance staff can perform the lubrications, adjustments, inspections, and other chores on a timely basis. Efficient, effective workers are still the key to effective service.

Contract Maintenance Services

As plants reduce their head count of employees, many companies are now entrusting conveyor maintenance to third-party specialist contractors. These contractors offer the breadth of expertise necessary to achieve high operating standards, and they allow plant maintenance staff to be re-deployed on other, perhaps more important projects in the plant.

The skilled contractor has the incentive to install a belt cleaner properly, at exactly the right position, the right cleaning angle, with the right blade-to-belt pressure. The contractor does not want to make unnecessary repeat visits to the plant, so he is under pressure to deliver the service required at minimum cost to both himself and his customer.

Contracting spillage cleanup to outside contractors helps a plant identify the true costs of the plant's inefficiency by forcing it to quantify what needs to be done to eliminate problems and to attach a cost to the remedial action.

Sometimes, spillage control is best treated as a joint exercise for both outside and in-house personnel. If it is left solely to the operating company, it may be neglected, as operators and maintenance personnel have too many other, more significant projects to devote their full attention to cleanup. On the other hand, if spillage control is contracted out totally, it is not healthy either. Under those circumstances, the in-house personnel will have no interest in operating the plant cleanly.

If a collaborative approach involving both in-house and outside personnel was chosen, benchmarks and performance standards could be identified within a contract based on the period of time over which improvements would become apparent. Some items, such as reduced cleanup time, would show an immediate improvement, while others, such as replacement costs on rollers, belts, and wear components, would only be discernible over a two-to-four-year period.

Recommended Conveyor Service

The specific service requirements for a conveyor will depend on both its construction and the volume and nature of materials moving through it. However, some basic rules apply that will help guide the formulation of a service schedule.

Belt

Make sure splices are properly installed and have no raised edges that can damage skirtboard seals and other components. Make sure there are no longitudinal rips or edge damage that should be repaired to preserve belt integrity and life.

Chute

Chutework should be examined to make sure there are no corrosion holes or missing bolts that will allow material to escape. The structural steel should provide rigid support for the panels.

Belt Support

Impact bars and support bars should be examined to ensure the belt has not worn through the top cover. Bars should be checked for delamination and straightness. Wear surfaces must be checked for abrasion, and bars turned over or replaced where necessary.

Pulleys

Make sure pulleys are aligned and turn evenly without slippage. They should be free of material buildup; remove any accumulated material.

Idlers

Rolling components should be inspected on a monthly basis and (if required) lubricated in accordance with manufacturer's specifications. Replace any roll that is not turning. Carrying idlers, return rolls, and snub pulleys should be cleaned periodically to prevent buildup and belt misalignment problems.

Wear Liners

Liners should be inspected for wear on the bottom edge. Make certain bottom edge is smooth and opens in direction of belt travel to prevent entrapment points. Check to make sure liners are securely installed, as a loose liner can slit a belt, and liners that fall onto the belt can damage a pulley or other component.

Skirtboard Seal

Each elastomer sealing strip should be inspected on a once-per-week or more frequent basis to make sure it "kisses" for effective containment of fines. Make certain to check the skirtboard seal when the belt is running and loaded, as the relationship of the belt to seal may be different when the conveyor is stopped or running unloaded.

Belt Cleaners

Cleaners and V-plows should be checked on a weekly basis, and blades replaced as required. Make certain the proper tension is applied to provide effective cleaning.

Belt Trackers

Belt steering systems should be free of material accumulation that can interfere with ability to function. Idler rolls and side guides should turn freely.

Dust Suppression

Lubricate the pump(s) as required. Check spray nozzles and filters for plugging due to dirt. Make certain chemical mixing equipment is working, and an adequate supply of the additive is in the system. If operating in a cold climate, make certain heating elements on plumbing systems are operational.

Dust Collectors

Make sure fans are operating as required. Check filters or collection media for condition. Make sure filter cleaning mechanisms are operating as specified. Check the pressure differential across the filters.

Safety Systems
> Check safety devices to make sure they are working, including belt wander switches and pull-rope emergency stop switches.

Periodic maintenance inspections, with replacement of equipment prior to the failure point, will ensure long life and minimal spillage. In addition, this "before necessary" maintenance philosophy will work to prevent a system failure and the loss of profits that a catastrophic downtime entails.

Idler Lubrication

As with all machinery, a well-developed program of lubrication is essential for low maintenance costs and dependable operations (unless a plant has standardized on idlers with sealed-for-life bearings). Because of the relatively large number of bearings in the idler rolls on a conveyor and their influence on belt tension and horsepower requirements, lubrication is very important. Following the idler manufacturer's recommendations as to the type, amount, and frequency of lubrication will enhance life expectancy.

Care should be taken not to over-lubricate idlers. This can damage the bearing seals, allowing fugitive material to enter the bearing and increasing the friction while decreasing idler life. Excess oil or grease can also spill onto the belt where it can attack the cover, decreasing the service life of the belt. Excess grease can also fall onto handrails, walkways, or floors, making them slippery and hazardous. Do not lubricate idlers equipped with sealed bearings.

Manuals

Make certain to check the owner/operator manuals issued by the supplier of any piece of equipment for specific instructions on service requirements, procedures, and timetables. A comprehensive file of equipment manuals should be kept and should be accessible to workers on all shifts.

In addition, maintenance personnel should keep careful records of inspections and service performed. This will ensure the proper maintenance of the equipment as well as validate any warranty claims against component manufacturers.

Spare Parts

Certain repair parts should be available in inventory. This will allow for both routine replacement of worn parts and speedy completion of unexpected repairs, allowing a faster return to operations. This inventory should include replacement "wear" parts and parts "likely to be damaged" such as belt cleaner blades, impact bars, and idlers. Also included in the maintenance stores should be a supply of "rip repair" belt fasteners for emergency repairs.

By standardizing across the plant on the variables, the size of this stockpile of spare parts (and, therefore, the expense for these idle parts) can be minimized.

It may be a good idea to keep a maintenance "bone yard" where items removed from service can be stored and cannibalized for necessary replacement parts as needed. However, parts taken from used equipment needs to be thoroughly cleaned and inspected before re-use.

Conveyor Start-Up

A belt is like a new pair of shoes; it needs to be "broken in" gradually and carefully to avoid painful moments. Insufficient attention at conveyor start-up–either in the initial running after construction or following a maintenance outage–can lead to significant and costly damage.

Typically, conveyor operations do not require the attention of many personnel. This is generally one of the selling points in the selection of belt conveyors over the other forms of haulage. But it would be a mistake to start up a belt, especially on a new conveyor or on a line that has received extensive modifications, without the attention of extra personnel along its route. Spotters should be in place along the belt run where trouble might be anticipated or would be particularly costly. These observers should be equipped with "walkie-talkies" or telephones and be positioned near emergency shut-off switches.

A careful inspection prior to start-up should establish that there are no construction materials, tools, or structural components left where they can gouge or cut the belt as it begins to move.

Modern electrical control systems can incorporate computers and other automatic means for measuring performance and controlling such functions as weighing, mixing, blending, and material path. Sensors and other devices for indicating maintenance requirements and unsafe conditions should be given a thorough check and debugging during the conveyor's initial trials.

The belt should be run empty, slowly at first, and then at what is anticipated to be normal operating speed as a check for possible problems.

Efficient and effective maintenance lowers costs, not just of the maintenance department, but for the total operations. The goal is to provide quality work with minimal disruption to the production routine. That will produce benefits in operating efficiency, system availability, and, ultimately, the bottom line.

Chapter 20

The Human Factor

Much of this volume has dealt with the hardware systems required to provide total material control for belt conveying operations. The time has come to recognize there is more to the control of fugitive material from belt conveyors than the latest hardware. There is another factor, one that is more crucial to the achievement of this goal. That factor is the human element.

No matter how innovative, sophisticated, specialized, or foolproof the technology, its long-term performance is governed by the human element.

In many ways, a plant is the reflection of the thinking of management, operations, and maintenance personnel. If these groups see the plant as dirty, inefficient, unpleasant, and unsafe, it will be allowed to become that way. The attitudes become self-fulfilling prophecies.

To make truly beneficial and long-lasting improvement in the performance of hardware systems, it is critical that the thinking of plant personnel of all levels be changed. If plant personnel have been taught to expect higher performance levels, and if they enact the measures to achieve and maintain these levels, then they can hold their conveyors (and other systems) to these higher expectations.

It is this emphasis on the human factors that can take the elements of total material control as spelled out in this book into a living process. It is this living process that is required to make a noticeable and long-term difference in the efficiency of plants. What is required is more than a hardware fix; it is a change in attitude and behavior that allows the creation of a process to provide a real improvement.

The improved hardware systems for materials containment and dust management described in this volume are an important step in achieving total material control. It is these human factors that provide the key to success.

The Limitations of Hardware

The goal of total material control requires more than hardware to solve. The first step in reaching the solution is to recognize that new components are not the sole answer.

Many times, new components are installed to upgrade a conveyor system and generate improvements in material control and plant efficiency. In most cases, these systems perform as expected and provide discernible benefits. However, the solution to fugitive material does not stop with equipment installations, regardless of how well designed or expensive the new systems are.

As important to the long-term correction of these problems is a plant's commitment to providing answers for these problems. This commitment

> "No matter how innovative, sophisticated, specialized, or foolproof the technology, its long-term performance is governed by the human element."

must be put continuously into practice by plant management and workforce alike.

Commitment to Improvement

Contrary to supplier claims, there are very few "one-shot" solutions where the application of new equipment provides a permanent improvement in bulk solids handling. Effective material control requires a process for the continuing improvement of conveyor operations. This means the plant must never be "satisfied" with conditions or results. Even "satisfactory" results and conditions change under the stress of day-to-day operation, varying material conditions, and minimal or non-existent maintenance. The plant must continuously search for opportunities to reduce costs and gain efficiency.

At the senior management level and at the operations and maintenance levels–"where the rubber meets the road"–personnel must understand that improvement is a process. The long-term answer to total material control requires a continuing sequence of improvement: refining, adjusting, upgrading, "tweaking," and maintaining the plant's system. Included in the plant's systems are the materials, the equipment, and the people.

Management must demonstrate a commitment to solve problems, and employees must have the ability to expend resources–time, capital, equipment–on the development and maintenance of the solutions.

A plant must call on all its resources to develop this process and achieve its goal of continuous improvement. It takes commitment from everyone, from the senior plant (and corporate) management to the newest operations or maintenance employee.

A Report Card from Consultants

Plant employees sometimes "can't see the forest for the trees." That is, they are so busy doing their jobs, they are unable to see the problems (or the opportunities) in front of them. The day-to-day tasks take all their time and concentration, and they are unable to stop doing the job to see the benefits of the possible improvement.

Or there is another factor: the inertia or the status quo. Plant employees have become accustomed to a certain procedure, a certain style, a certain level of performance, certain circumstances, certain conditions, and even certain problems. They are unaware there is another way to do things, or that another result could be achieved.

This is where outside consultants can make a significant contribution to an operation. These engineers and material-handling specialists can analyze plant systems and procedures and provide a "report card."

Figure 20.1
Consultants will help plant personnel see the strengths and weaknesses of the operation's material handling system.

Based on a detailed observation of an operation, these consultants can assess the plant material handling systems, to point out places where improvements in equipment and practices would be beneficial. They provide an expert resource on what is possible and what other plants in a given industry or with similar equipment are achieving (**Figure 20.1**)

These consultants can help plant, operations, maintenance, and management personnel see the strengths and weaknesses of a material handling system. They can provide a yardstick for material handling system performance, appraising what is working well, and what does not meet industry standards. They can help prioritize improvements to an existing system, to provide a prompt return on investment.

Becoming Partners with Suppliers

One asset that may go under-utilized in the attempt to establish total material control is a plant's ability to develop a partnership with its suppliers. The plant must take advantage of the knowledge base available from these vendors. Salesmen–and their corporate resources, like applications experts, industry managers, product engineers, and installation specialists–can make important contributions to a plant's program for control of fugitive material and the improvement of operations. The key is for suppliers to become partners with the plant.

So how does a supplier become a partner? A partner will tell you what you are doing right and what you are doing wrong. A partner will tell you when equipment

Figure 20.2

By making an important contribution to a plant's success, suppliers move from vendors to partners.

alone will not solve the problem and what attributes are missing. A partner will share in the risk and the reward. Most importantly, a partner earns the right to be there, with energy, effort, and results.

There can be no "get the purchase order and get out of town" mentality for suppliers and salesmen. The key to success for both supplier and plant is to establish a long-term relationship centered on improving efficiency and meeting needs. The most essential ingredient of this process is open and truthful communication between vendor and plant, sales representative and customer. **(Figure 20.2)**

Training

One important aspect of the drive for continuing improvement in material control is training. Employees must be taught what to expect from equipment and what is required. They should learn how to identify problems, and then how to troubleshoot and adjust the systems to correct and minimize these problems.

All available information sources must be integrated into a knowledge base that can train new hires and further educate existing personnel. Consultants can offer training on the characteristics of bulk solids and equipment performance. Suppliers can and must provide information on the installation, maintenance, and troubleshooting of their systems. Veteran employees can provide information on the in-plant process, the key procedures, and the likely problems.

Figure 20.3

Maintenance must be an integral part of a plant's continuous improvement process.

Just as consulting engineers and suppliers must be educated about the specific characteristics of plant and material, the plant personnel must be trained on the capabilities of the new system. They also need to be trained in effective maintenance procedures to keep system effectiveness high.

The Importance of Service

As discussed elsewhere in this volume, maintenance is a critical function in assuring efficient conveyor operation. Consequently, maintenance must become an integral part of a plant's process of continuous improvement. **(Figure 20.3)**

In most cases, the systems for total material control require only minor adjustment, not sophisticated procedures. But when these procedures are not performed, small problems can turn into big trouble.

But with increasing demands on plant maintenance crews that may already be "reduced-in-force" and overtaxed, it becomes more important to find alternative solutions. It is now more accepted to call in outside contractors to perform even routine maintenance tasks.

Outsourcing the basic procedures is one way to provide for and control the costs of routine service. These specialists improve the speed, efficiency, and regularity of routine maintenance. They are experts in maintenance for a particular set of components, so they improve the quality and speed of service operations. A skilled contractor has technical knowledge about his specialties, and has an incentive to optimize his chores–for example, to install a belt cleaner properly–at exactly the right position, the right cleaning angle, with the right blade-to-belt pressure. The contractor does not want to make unscheduled repeat visits–callbacks–to the plant, so he is under pressure both from his customer and from himself to deliver the service required for optimal performance at minimum cost.

These equipment service specialists also can assist in the specialized training required to keep plant maintenance personnel up to speed on more complicated procedures and new technologies.

The use of these contracted specialists allows plant maintenance staff to focus less on routine maintenance and more on planning, less on solving minor troubles and more on major projects.

Regardless of whether the service is performed by outside contractors or in-house personnel, a commitment to the maintenance of material control systems is essential to their overall improvement.

Development of a Process

Reducing fugitive material and improving conveyor performance is not a question of buying the latest piece of hardware from the next peddler to walk through the front door. Instead, the solution lies in developing a process that can continuously maintain and upgrade performance–the performance of suppliers, equipment, materials, and personnel. The process begins with a plant-wide commitment to improving management of fugitive material. It features the development of relationships with suppliers and consultants who can assist the plant in focusing on the problem. It includes optimization of service to maintain component performance and improve overall plant operating efficiency. These steps help develop a program of continuing refinement in pursuit of the elusive goal of total material control.

This process of continuing improvement is not developed overnight. Nor can it be accomplished by decree from high up the chain of command. While it does require the endorsement of senior officials, it equally requires the commitment of operations and maintenance personnel. This process must be carefully nurtured through the learning curve of education and experience.

The Value of Supplier Expertise

The trend in industry is to reduce the number of full-time permanent employees and rely on outside sources of labor and expertise. Many start-up companies are looking at this as an opportunity to provide a wide range of services.

In the past there has been great reluctance for companies to enter into long-term relationships with suppliers, as the thinking was that the supplier would only do what was good for the supplier, and not for the customer. This is no longer true. Now, an open relationship with key suppliers can be a major competitive advantage. The business environment has changed dramatically and there have been many attempts to adapt the supplier base to these changes.

Buying groups and preferred suppliers are ways companies are trying to reduce bureaucracy and increase participation by suppliers. Often these concepts have failed because they failed to recognize the expertise of the supplier and the dependency of plant personnel on this supplier experience and capability. Similarly, too often project work has been contracted to installation and maintenance companies that do not specialize, only generalize.

Total Material Control is a concept that requires a great deal of specialization and attention to detail. When choosing a consultant or service company to assist in achieving Total Material Control careful consideration must be given to the supplier's core competency and ability to fulfill the commitments made. The ability of the supplier to engineer situation-specific solutions, to manufacture and install them, and then to maintain them to a single performance standard is a strategic advantage worth seeking.

Total Material Control

There are a number of hardware solutions aimed at total material control. But the key ingredient is the "software." It is "the human factor" that will determine its eventual success or failure.

Acknowledgments

This book could not have been completed without the understanding and assistance of many Martin Engineering employees. These individuals have provided background information, technical expertise, "big picture" thinking, and "nuts-and-bolts" details. The list (given here in alphabetical order) includes, but is not limited to, the following:

Jim Burkhart, Kevin Collinson, Kim Frank, Steve Frank, Daniel Marshall, Fred McRae, Dave Mueller, Jared Piacenti, Mark Stern, Gary Swearingen, Dennis Taylor, Larry Thomas, Andy Waters, and Marty Yepsen.

Martin Engineering consultants and specialists Paul Grisley, Dick Stahura, and Willem Veenhof provided a significant contribution in their respective areas of expertise.

Lending a long-distance hand were Jorg Gauss of Martin Engineering GmbH, Walluf, Germany, and Bob Law, Managing Director of Engineering Services and Supplies, the Martin Engineering licensee in Australia.

Jean-Francois Schick of GORO, S.A. provided a timely review of Chapter 3. GORO S.A. also provided illustrations for use in the discussion of belt splicing. We obtained illustrations from the Dust Collection Group of Donaldson Company, Inc., through the good efforts of Accounts Manager Tom Krippner.

We appreciate Martin Engineering administrative assistants Jill Kelman, Marian Pletkovich, Marianne Pyle, and Bonnie Thompson for their care and diligence in proofreading. It is always a pleasure to work with Kewanee, Illinois' leading retired newspaperman/freelance proofreader Dennis Sullivan.

Becky Kubiak, Bob Tellier, and Tracey Swanson, the designers in Martin Engineering's Marketing Department, developed the format, compiled the illustrations, and fit all the pieces together. They remained cool despite late additions and last minute-corrections, and put it all together in this pleasing and usable package.

To all who lent a hand, thank you.

RTS, LJG, ADM
Neponset, Illinois
February 2002

References

Arnold, P. C., "Transfer Chutes Engineered for Reliable Performance," Paper presented at The Institution of Engineers, Australia, 1993 Bulk Materials Handling National Conference, Queensland, Australia, September 1993, National Conference Publication No. 93/8, p. 165-173.

Axelrod, Steve, "Maintaining Conveyor Systems," *Plant Engineering*, Cahners Publishing Company, Des Plaines, Illinois, September 1994, p. 56-58.

Barfoot, G.J., "Quantifying the Effect of Idler Misalignment on Belt Conveyor Tracking," *Bulk Solids Handling*, Trans Tech Publications, Clausthal-Zellerfeld, Germany, January/March 1995, p. 33-35.

Belt Conveyors for Bulk Materials, Conveyor Equipment Manufacturers Association (CEMA), Fifth Edition, 1997.

Benjamin, C.W., "Transfer Chute Design: A New Approach Using 3D Parametric Modelling," *Bulk Solids Handling*, Trans Tech Publications, Clausthal-Zellerfeld, Germany, Jan/March 1999, p. 29-33.

Bennett, D.J., "Improve Conveyor Performance Through Effective Monitoring of Belt Cleaning Systems," Paper presented at the IIR Conference *Improving the Performance of Conveyor Operations in Mining*, Sydney, Australia, August 1996.

Care and Maintenance of Conveyor and Elevator Belting, The B.F. Goodrich Company, Akron, Ohio, 1980.

Carter, Russell A., "Knocking Down Dust," *Rock Products*, Intertec Publishing, Chicago, May 1995, p. 19-23, 40-44.

CEMA Standard No. 575-2000., "Bulk Material Belt Conveyor Impact Bed/Cradle Selection and Dimensions," Conveyor Equipment Manufacturer's Association (CEMA), Naples, FL, 2000.

Colijn, Hendrik, *Mechanical Conveyors for Bulk Solids*, Elesevier Science Publishers B.V., Amsterdam, The Netherlands, 1985.

Conveyor Terms and Definitions, Conveyor Equipment Manufacturers' Association (CEMA), Rockville, Maryland, Fifth Edition, 1988.

Cooper, Paul, and Tony Smithers, "Air Entrainment and Dust Generation from Falling Streams of Bulk Materials," Paper presented at 5th International Conference on Bulk Material Storage, Handling and Transportation, Wollongong, Australia, July 1995.

Cukor, Clar, *Tracking: A Monograph*, Georgia Duck and Cordage Mill, Scottdale, Georgia, (undated).

Drake, Bob, "Cures for the Common Pulley," *Rock Products*, Intertec Publishing, Chicago, May 2001, p. 22-28.

Finnegan, K., "Selecting Plate-Type Belt Fastener Systems for Heavy-Duty Conveyor Belt Operations," *Bulk Solids Handling*, Trans Tech Publications, Clausthal-Zellerfeld, Germany, May/June 2001, p. 315-319.

Fish, K.A., A.G. Mclean, and A. Basu, "Design and Optimisation of Materials Handling Dust Control Systems," Paper presented at the 4th International Conference on Bulk Materials Storage, Handling and Transportation, Wollongong, Australia, July 1992.

Friedrich, A.J., "Repairing Conveyor Belting Without Vulcanizing," in *Bulk Material Handling by Conveyor Belt III*, SME, Littleton, Colorado, 2000, p. 79-85.

Gibor, M., "Dust Collection as Applied to Mining and Allied Industry," *Bulk Solids Handling*, Trans Tech Publications, Clausthal-Zellerfeld, Germany, July/September 1997, p. 397-403.

Godbey, Thomas, "Dust control systems: Make a wise decision," *Chemical Processing*, Putnam Publishing, Chicago, May 1990, p. 23-32.

Godbey, Thomas, "Selecting a dust control system (Part I)," *Powder and Bulk Engineering*, CSC Publishing, Minneapolis, October 1989, p. 37-42.

Godbey, Thomas, "Selecting a dust control system (Part II)," *Powder and Bulk Engineering*, CSC Publishing, Minneapolis, November 1989, p. 20-30.

Goldbeck, Larry J., Martin Engineering, "Controlling fugitive material at your belt conveyor's loading zone," *Powder and Bulk Engineering*, CSC Publishing, Minneapolis, July 1988, p. 40-42.

Goldbeck, Larry J., "Matching Belt Compatibility to Structures," *Aggregates Manager*, Mercor Media, Chicago, July 2001, p. 21-23.

Greer, Charles N., "Operating Conveyors in the Real World," *Rock Products*, Maclean-Hunter Publications, Chicago, April 1994, p. 45-48.

Grisley, Paul, "Air Supported Conveying in Mines and Process Plants," Paper presented at the 2002 SME Annual Meeting & Exhibit, Phoenix, AZ, February 2002.

Handbook of Conveyor & Elevator Belting on CD, Version 1.0, Goodyear Tire & Rubber Company, Akron, Ohio, 2000.

"Hints & Helps: Tips for Tracking Conveyor Belts," *Rock Products*, Intertec Publishing, Chicago, February 1995, p. 25.

Industrial Ventilation: A Manual of Recommended Practice, American Conference of Governmental Industrial Hygienists, 22nd Edition, Cincinnati, OH, 1995.

Kasturi, T.S., *Conveyor Belt Cleaning Mechanism*, Jay Kay Engineers & Consultants, Madras, India, 1992.

Kasturi, T.S., "Conveyor Belting Wear: A Critical Study," Unpublished study commissioned by Martin Engineering, Jay Kay Engineers & Consultants Madras, India, May 1995.

Kasturi, T.S., *Conveyor Components, Operation, Maintenance*, Failure Analysis, Jay Kay Engineers & Consultants, Madras, India, 1994.

Kestner, Dr. Mark, "Using suppressants to control dust emissions (Part I)," *Powder and Bulk Engineering*, CSC Publishing, Minneapolis, February 1989, p. 17-20.

Kestner, Dr. Mark, "Using suppressants to control dust emissions (Part II)," *Powder and Bulk Engineering*, CSC Publishing, Minneapolis, March 1989, p. 17-19.

Koski, John A., "Belt conveyor maintenance basics," *Concrete Journal*, The Aberdeen Group, Addison, Illinois, March 1994, p. 5.

Law, Bob, "Conveyor Belt Cleaner Analysis," Paper presented at the IIR Conference *Improving Conveyor Performance*, Perth, Australia, August 2000.

Low, Allison, and Michael Verran, "Physical Modelling of Transfer Chutes–A Practical Tool for Optimising Conveyor Performance," Paper presented at the IIR Conference *Improving Conveyor Performance*, Perth, Australia, August 2000.

Miller, D., "Profit from Preventive Maintenance," *Bulk Solids Handling*, Trans Tech Publications, Clausthal-Zellerfeld, Germany, January/March 2000, p. 57-61.

Mody, Vinit, and Raj Jakhete, *Dust Control Handbook*, Noyes Data Corporation, Park Ridge, New Jersey, 1988.

Möller, J.J., "Protect Your Conveyor Belt Investment," Presentation to BELTCON 3 International Material Handling Conference, Johannesburg, South Africa, September 1985.

Morgan, Lee, and Mike Walters, "Understanding your dust: Six steps to better dust collection," *Powder and Bulk Engineering*, CSC Publishing, Minneapolis, October 1998, p. 53-65.

Muellemann, Alf, "Controlling dust at material transfer points with ultra-fine water drops," *Powder and Bulk Engineering International*, CSC Publishing, Minneapolis, January 2000, p. 44-47.

NIBA Engineering Handbook, National Industrial Belting Association, Brookfield, WI, 1985.

Öberg, Ola, "Materialspill vid bandtransportörer" (Material Spillage at Belt Conveyors), Royal Institute of Technology, Stockholm, Sweden, 1986.

Ottosson, Goran, "The cost and measurement of spills and leaks at conveyor transfer points," *World Cement Materials Handling Review*, Berkshire, England, October 1991.

Reed, Alan R, "Contrasting National and Legislative Proposals on Dust Control and Quantifying the Costs of Dust and Spillage in Bulk Handling

Terminals," *Port Technology International*, ICG Publishing Ltd., London, 1995, p 85-88.

Rhoades, C.A., T.L. Hebble, and S.G. Grannes, *Basic Parameters of Conveyor Belt Cleaning*, Bureau of Mines Report of Investigations 9221, US Department of the Interior, Washington, D.C., 1989.

Sabina, William E., Richard P. Stahura, and R. Todd Swinderman, *Conveyor Transfer Stations Problems and Solutions*, Martin Engineering Company, Neponset, Illinois, 1984.

Scott, Owen, "Design Of Belt Conveyor Transfer Stations For The Mining Industry," Proceedings of the 1993 Powder & Bulk Solids Conference, Reed Exhibition Companies, Des Plaines, Illinois, p. 241-255.

Simpson, G.C., "Ergonomics as an aid to loss prevention," *MinTech '89: The Annual Review of International Mining Technology and Development*, Sterling Publications Ltd., London, 1989, p. 270-272.

Stahura, Dick, Martin Engineering, "Ten commandments for controlling spillage at belt conveyor loading zones," *Powder and Bulk Engineering*, CSC Publishing, Minneapolis, July 1990, p. 24-30.

Stahura, Richard P., Martin Engineering, "Conveyor skirting can cut costs," *Coal Mining*, McLean-Hunter Publications, Chicago, February 1985, p. 44-48.

Sundstrom, P., and C.W. Benjamin, "Innovations In Transfer Chute Design," Paper presented at the 1993 Bulk Materials Handling National Conference, The Institution of Engineers, Australia, Conference Publication No. 93/8, p. 191-195.

Swinderman, R. Todd, and Douglas Lindstrom, Martin Engineering, "Belt Cleaner and Belt Top Cover Wear," Paper presented at the 1993 Bulk Materials Handling National Conference, The Institution of Engineers, Australia, National Conference Publication No. 93/8, p. 609-611.

Swinderman, R. Todd, Martin Engineering, "Belt Cleaners, Skirting and Belt Top Cover Wear," *Bulk Solids Handling*, Trans Tech Publications, Clausthal-Zellerfeld, Germany, October-December 1995.

Swinderman, R. Todd, Martin Engineering, "Conveyor Belt Impact Cradles: Standards and Practices," Paper presented at the 2002 SME Annual Meeting & Exhibit, Phoenix, AZ, February 2002.

Swinderman, R. Todd, Martin Engineering, "Engineering your belt conveyor transfer point," *Powder and Bulk Engineering*, CSC Publishing, Minneapolis, July 1994, p. 43-49.

Swinderman, R. Todd, Steven L. Becker, Larry J. Goldbeck, Richard P. Stahura, and Andrew D. Marti, *Foundations: Principles of Belt Conveyor Transfer Point Design and Construction*, Martin Engineering, Neponset, Illinois, 1991.

Swinderman, R. Todd, Larry J. Goldbeck, Richard P. Stahura, and Andrew D. Marti, *Foundations^2: The Pyramid Approach to Control Dust and Spillage From Belt Conveyors*, Martin Engineering, Neponset, Illinois, 1997.

Swinderman, R. Todd, Martin Engineering, "The Conveyor Drive Power Consumption of Belt Cleaners," *Bulk Solids Handling*, Trans Tech Publications, Clausthal-Zellerfeld, Germany, May 1991, p. 487-490.

Thomas, Larry R., Martin Engineering, "Transfer Point Sealing Systems To Control Fugitive Material," 1993 Bulk Materials Handling National Conference of The Institution of Engineers, Australia, Conference Publication No. 93/8, p. 185-189.

Tostengard, Gilmore, "Good maintenance management," *Mining Magazine*, The Mining Journal, Ltd., London, February 1994, p. 69-74.

Weakly, L. Alan, "Passive Enclosure Dust Control System," in *Bulk Material Handling by Conveyor Belt III*, SME, Littleton, CO, 2000, p. 107-112.

Wilkinson, H.N., Dr. A.R. Reed, and Dr. H. Wright, "The Cost to UK Industry of Dust, Mess and Spillage in Bulk Materials Handling Plants." *Bulk Solids Handling*, Trans Tech Publications, Clausthal-Zellerfeld, Germany, Volume 9 Number 1, February 1989, p. 93-97.

Index

A-frame skirtboard supports, 91
access, 196-199
   doors, 198-199
   to belt cleaners, 136
   to chutes, 59, 198-199
   to skirtboard seals, 91
accidents, 192-195
   cost of, 3-4
angle of repose, 38-39
air cannons on chutes, 59
air movement, 108-113, 128-129
air-supported conveyors, 10-11, 68
atomization, 116-118

baghouse dust collectors, 124-126
belt carcass, 16-17
belt covers, 17
belt damage, 21-27, 60-61, 162-163, 202-203
belt feeders, 38-39
belt repair, 25-27
   adhesives, 26
   mechanical fasteners, 27
belt sag, 61-62
belt speed monitors, 202
belt tension, 19
belt training, procedure for, 172-173
belt turnovers, 156-157
belt washing stations, 148-149
belt wear
   and belt cleaners, 105
   and sealing systems, 151
belting
   repair of, 25-27
   storage and handling of, 20-21
benchmarks for conveyor surveying, 171
bend radius, 19
blade-to-belt pressure, 141

brush cleaners, 148
Bureau of Mines belt cleaning research, 141, 146

cable belt conveyors, 11
camber, 25
capacity, belt,
   and edge sealing, 100-101
   and trough angles, 39-40, 43, 87-88
CARP, see constant angle belt cleaners
carryback, 1-2, 132-137
cartridge filter dust collectors, 127
cascade transfer, 52
catenary idlers, 66-67
catenary idler stabilizer, 67-68
central dust collection systems, 125
centralizers, load, 49-50
ceramic liners, 57-58, 94-97
chatter, belt cleaner blade, 136, 138, 139, 140
chevron belt cleaner, 147
chevron belts, 11-12, 147
chute modeling, 55
chute width, 56, 87-88
cleaning pressure, 141
cleated belts, 11-12, 147
computer modeling, 55
constant angle belt cleaners, 140
contract maintenance services, 203-204, 208-209
cradles, 69-83
   impact, 72-74
   seal support, 69-70
creep drive, 202
crowned pulleys, 176-177
cut edge belts, 18
cyclones, 126-137

deflected blade cleaners, 141-142, 146

deflectors, 47-48
deflector wear liner, 93
diagonal plows, 158-161
displaced air, 110
doors
    access, 136, 198-199
    leakage through, 54-55
    tail gate, 45
"double height" skirtboard, 88-89
dressing a splice, 35
dribble chutes, 155-156
    vibrating, 156
dust bags, 113
dust curtains, 111-112
dust explosions, 129-130

edge damage, 23
edge distance, 87-88, 104
edge seal support, 69-70
effective belt width, 87-88
electrostatic precipitators, 127-128
enclosed roller conveyors, 12
entrapment, 22-24, 60-62, 69-70, 89-91, 96-97
entry area sealing, 43-45, 112-113
ergonomics, 201, 206-209
exhaust air requirements, 108-113, 128-129
exit area sealing, 111-112
explosions, dust, 129-130

feeder belts, 38-39
filter materials, 128
filter relief bags, 113
flat belts, 38-39
floating skirting seals, 103-104
flow training gates, 50
foam suppression systems, 119-121
fog dust suppression systems, 116-118
food-grade belt cleaners, 147
footings, conveyor structure, 60
full-trough pulleys, 41

generated air, 111
grizzly bars, 52
grooves in belt, 50, 61, 105-106
guards on equipment, 195

half-trough pulleys, 41, 90
"heeling," 152
"hood and spoon" chutes, 55-56, 71, 111
horizontal-curved conveyors, 12

horsepower requirements, see power consumption

ICT carryback gauge, 134
idler lubrication, 205
idlers, 62-68, 71-72
    catenary, 66-67
    closely-spaced, 64-65
    impact, 71-72
    intermediate, 76
    picking, 63
    return, 63-64
    self-aligning, 64, 177-181
    tilted, 171-172
    training, 64, 177-181
    transition idlers, 39-41, 79-80
    V-return, 64
impact, 50-53, 70-74
impact bars, 72-73
impact cradles, 71-74
impact damage, 22, 70-74
impact grids, 53
impact plates, 53
induced air, 52, 109-110
insertable dust collectors, 126
inspections, 202
integrated dust collection systems, 126
intermediate idlers, 76
ISO 14000, 5

"knocking" idlers, 173-174

laser surveying, 167-172
liners, chute, 547-58
load centralizers, 49
loading zone impact, 66-67, 70-75
lockout/tagout, 172, 193-194
lubrication, idler, 205

magnetic pulleys, 202-203
magnets, suspended, 202-203
mandrel belt cleaner mount, 153
manuals, 205
mechanical splices, 33-36
misalignment switches, 175-176, 202
mistracking, 162-181
    and skirt seals, 105
    on reversing belts, 165, 180-181
molded-edge belts, 18
monitors, belt speed, 202
multiple cleaner systems, 143-146
multiple-layer sealing systems, 102-103
multiple loading points, 91

observation requirements, 198
off-center loading, 47-50, 164-166

parts, spare, 205
passive dust collection, 113
peeling angle for belt cleaning, 139
pentaprism, 167-170
plugged chute switches, 202
pneumatic belt cleaners, 148
pocket conveyors, 12-13
"pooling," 53-54
power consumption
    and belt cleaners, 149-151
    and belt support systems, 77-83
    and belt trackers, 181
    and skirtboard seal, 104-105
precipitators, electrostatic, 127-128
pre-cleaners, 144-45, 152
pulley cleaners, 156
pulleys
    crowned, 176-177
    magnetic, 202
    take-up, 9, 162, 168-170
    wing, 37-38, 44
    wrapped, 37-38, 44

repair of belting, 25-27
residual suppression, 121
reversing belt cleaners, 147
rip detectors, 203
rip repair fasteners, 27
rock boxes, 52-53
rock ladders, 52
rollback, material, 43-45

scab plates, 55
scavenger conveyors, 156
scraping angle for belt cleaning, 139
sealing systems
    edge, 98-107
    entry, 43-45, 112-113
    exit, 111-112
    maintenance of, 106, 204
secondary cleaners, 145-146
skiving, 35
spacing
    conveyor, 196-198
    idler, 64-65
spare parts, 205
speed-up belts, 52-53
split rubber curtains, 111-112
"spoon" chutes, 48, 55
Stahura Carryback Gauge, 134

start-up, 174, 205
steep-angle conveyors, 13
stilling zone, 112
straight wear liner, 93
surfactants, 118-121
surveying of conveyors, 167-171
switches
   emergency stop, 194
   misalignment, 175-176, 202
   plugged chute, 202

tail gate sealing box, 43-45
tail protection plows, 158-161
telescoping belt cleaners, 153
television, closed circuit, 203
tensioner, belt cleaner, 141-142
thermal penalty for moisture addition, 115-116
tilted idlers, 174
training gates, material flow, 50
training idlers, 63-64, 177-181
training the belt, procedure for, 172-173

training, personnel, 193, 208-209
tramp iron, 23, 25, 202-203
tramp iron detectors, 202-203
transition, 20, 39-42
   distance, 20, 40-41
   gradual, 42
   idlers, 41
   two-stage, 42
   use of cradles in, 42
transition belts, 52-53
trough angles, 19-20, 38-39
   and belt capacity, 87-88
tube conveyors, 13-14
turnovers, belt, 156-157

UHMW Polyethylene
   as a liner, 57-58, 95-96
   in sealing support bars, 72
unit dust collection systems, 125-126
urethane liners, 95-96
used belting, 18, 100, 138

V-plows, 158-161
V-return idlers, 63, 176
vertical edge guides, 176
vibrating dribble chutes, 157-158
vibrating scavenger conveyor, 156
vibrations
   belt, and belt support systems, 60-61, 69-77
      from wing pulleys, 37-38, 44
      impact on belt cleaning, 136
vibrator on a chute, 58
vulcanized splices, 28-30

walkways, 197
wash box belt cleaners, 148-149
water spray, 114-116, 148-149
wear liners, 92-97
wet scrubbers, 127-128
wing pulleys, 37-38, 44
"worst case" material conditions, 51, 136-137

FOUNDATIONS Workshop
Operation and Maintenance of Clean and Safe Belt Conveyors

Schedule a workshop on how to prevent dust and spillage and improve conveyor operations in your operation.

Martin Engineering's non-commercial FOUNDATIONS Workshop will provide practical information that will improve your material handling.

MARTIN ENGINEERING
One Martin Place
Neponset, IL 61345 USA

For more information,
Phone 309-594-2384 Ext. 331
or 1-800-544-2947 Ext. 331
Fax 309-594-2432
E-Mail sueg@martin-eng.com
Or visit: www.martin-eng.com

Want to Know More?

- Knowledgeable Leader; Lively Format.
- Presented at Your Plant, or Company Headquarters, a Neutral Site, or a Regional Association Meeting.
- Trains Personnel from Newest Operator to Senior Maintenance Foreman; Engineers and Managers Too.
- Continuing Education Units (CEUs) Available.
- Half-Day or Full-Day Formats.
- Can Include Survey and Discussion of Problems at Your Plant (Requires Prior Arrangement).

Workshop Topics:

- Belt Damage: Causes And Cures
- What to do about Belt Wander
- The Importance of Wear Liner
- How to Size Dust Collectors
- Suppression and Collection: How They Can Work Together; When They Won't...
- The Cost of Fugitive Material
- Controlling Belt Sag
- How to Prevent Entry Leakage
- Plus Problems in Your Operation